Management Guide for CIM

Computer-Integrated Manufacturing

Management Teams Report Steps for Successful Business Strategies

Complimentary Copy
provided by:

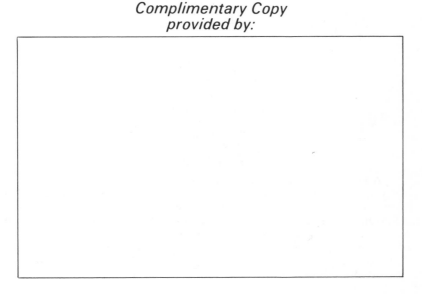

Edited by
Nathan A. Chiantella
IBM Corporation

Published by
The Computer and Automated Systems
Association of SME
One SME Drive
P.O. Box 930
Dearborn, Michigan 48121

Management Guide for CIM
Computer-Integrated Manufacturing
Management Teams Report Steps for
Successful Business Strategies

Copyright 1986
The Computer and Automated Systems
Association of SME
Dearborn, Michigan 48121

First Edition

Fourth Printing

Library of Congress Catalog Card Number: 86-70961
International Standard Book Number: 0-87263-241-5

PREFACE

The Management/CIM Challenge

Computer-Integrated Manufacturing, CIM, is the most exciting thing happening in the goods producing sector in recent times.

Accompanying the opportunities of CIM is the new management challenge of how to make it happen within appropriate segments of a company, division, plant, and product group. This problem surfaces because CIM by its very nature requires a unique collaborative effort by the entire management team involved.

The management team considering CIM must learn to focus collectively on a single system view across the entire business cycle of manufacturing. Hence, CIM introduces new management roles, techniques and actions.

Where do interested industry practitioners turn to become informed on "how-to" manage the CIM opportunity? The Society of Manufacturing Engineers has instituted this project in order to produce a management guide containing the most current information available.

Purpose of this Guide

This guide is intended for management in the goods producing sector. Its purpose is to bring together widely dispersed information on the new, emerging CIM-driven manufacturing strategy. The object of this publication is to guide management through the key phases of a CIM program. The phases included span from management's initial general interest in CIM to installed operations.

This guide is written by managers from companies that are successfully utilizing CIM. The authors emphasize the key role management must play. The guide details these activities in an organized and sequential manner. The focus is on management techniques and actions required to realize CIM opportunities. The technical composition of CIM implementation technology, products, and vendors are not within the scope of this guide.

The authors are from many types of companies, have different responsibilities and do not share common business requirements. Their CIM solutions are also unique.

But their management techniques used to create these solutions are sufficiently common to make them extremely valuable.

To capture and relate these individual experiences, the guide is structured to follow the sequential phases of a generic CIM cycle. Each phase contains several articles describing management's role and applied techniques. In this manner, the reader will follow sequential information leading from the start to the end of the full CIM cycle of management experience.

Each article contains the author's name and company.

Organization of the Guide by Topic

To provide the reader with sequential "how-to" management information, the guide is divided into topic categories. Each topic covers a sequential step leading through the management of the CIM cycle. A listing of categories is provided followed by detailed articles on each topic.

Categories of the Guide

How to:
- View CIM as a Business Strategy
- Overview Management's Role in CIM
- Spark Initial Interest in CIM
- Develop a Climate for a CIM Initiative
- Form a Team to Seek CIM Opportunities
- Conceptualize the New Way Using CIM
- Structure a CIM Program Proposal
- Sell the CIM Proposal
- Manage the Implementation Phases of CIM
- Measure and Evaluate CIM Results
- Move to the Next CIM Opportunity
- Summarize and Share CIM Experiences

Nathan A. Chiantella
Editor

IN APPRECIATION

The content of this publication was developed under the auspices of SME by industry professionals and without compensation.

The authors, with the support of their company management, gave freely to share hard-earned CIM experience. They have transformed vague concepts into practical guidelines for managers to pursue a CIM business/product strategy. SME appreciates the efforts of the contributing authors who are identified with each article.

In addition, there are many others who supported the idea to create this guide. The following is a partial list of those who made their company resources available to help produce this publication.

Randall J. Alkema
General Electric Company

Paul W. Brauninger
Cone Drive Operations, Ex-Cell-O Corporation

George J. Hess
The Ingersoll Milling Machine Company

Fred Houtz & Donald Sage
AT&T Technologies, Inc.

Jerry L. Latta & James J. Stifler
IBM Corporation

TABLE OF CONTENTS

STAGE 4 IMPLEMENTATION OF CIM

Stage 1
Introduction to CIM

Stage 2
Preparation for CIM

Stage 3
Program Plan for CIM

Stage 4
Implementation of CIM

A CIM BUSINESS STRATEGY FOR INDUSTRIAL LEADERSHIP

Nathan A. Chiantella
IBM Corporation

Transformation for Leadership

The goods producing sector is undergoing examination throughout the industrialized world. The motivation is a race to discover the secrets behind a major transformation which is underway. The information being sought will be vital to secure industrial leadership. Management effectiveness will determine how rapidly this information will be obtained and utilized to improve the competitiveness of their products and industrial complex.

In the years ahead competition will intensify. Environmental conditions in the manufactured goods sector are changing. The trend is away from the narrow perspective of merely adjusting production capacity to fill demand. Today, industry is aggressively seeking a total competitiveness across the entire cycle of product design, manufacture, and marketing. The quest for higher efficiencies and faster cycle times as a collective business unit is straining traditional relationships among business headquarters, labs, plants, and sales branches.

The manufacturing industry is evolving from its well understood caterpillar stage and has not yet emerged as some new species of butterfly. The management challenge is to set a viable business strategy for this transformation.

CIM...A Step Into the Future

This is indeed a difficult period of preparation for the next major industrial evolution. Yesterday's strategies, practices, and experience have carried industry successfully to the door of the future world of industrially produced products. It is evident that major change lies ahead.

There are no clear signals on which road to travel. Some industry leaders have partially entered the future. They report that CIM, Computer-integrated manufacturing, is a viable long-term direction for total competitiveness across the entire product cycle. The benefits achieved show that CIM provides significantly larger gains than traditional methods as reflected in *Table 1*.

TABLE 1	
The Benefits of CIM*	
Reduction in engineering design cost	15-30%
Reduction in overall lead time	30-60%
Increased product quality as measured by yield of acceptable product	2-5 times previous level
Increased capability of engineers as measured by extent and depth of analysis in same or less time	3-35 times
Increased productivity of production operations (complete assemblies)	40-70%
Increased productivity (operating time) of capital equipment	2-3 times
Reduction of work in process	30-60%
Reduction of personnel costs	5-20%

*The companies studied expect further benefits as full integration is approached.

Source: Computer Integration of Engineering Design and Production
Manufacturing Studies Board
National Research Council, Washington, D.C.

Management's Leadership Role in CIM

Some new terms and conditions are required in order to enter the arena of CIM-size benefits. For one thing, these proposals for improvement do not travel established corporate routes. CIM program proposals do not originate from any particular part of an organization. And, CIM proposals do not flow upwards to solicit the usual management reviews for funding. This is because CIM is not achieved by looking for factory fixes within a financial culture which expects to justify improvements by direct labor cost reductions. Those who are successful say that CIM requires management to become active initiators and participants.

The CIM opportunity presents both management and technological challenges. These are complementary and interdependent activities. The combined requirements are to create a CIM-based business strategy to manage the entire product cycle and to leverage modern technology for strategy implementation. Because CIM couples management leadership with technological leadership, maximum competitiveness is addressed.

The scope of this writing is to concentrate on the management aspects of the CIM opportunity and not the technological implementations. The goal is to provide information to enhance management's role in creating the corporate environment in which CIM technological leadership will flourish.

Management's Introduction to CIM

For management, the initial excursion into the exciting new world of CIM is likely to be a rewarding as well as a frustrating experience. It is rewarding to visit a CIM installation and watch it in operation. The local management explanations of the new CIM principles by which it operates are easy to follow and understand. The briefings on the benefits achieved will generate envy. One is motivated to initiate a CIM program upon arrival back at home base.

Having returned, the rewarding portion of the experience fades and is displaced by management frustration. The initial problem management faces is the lack of a methodology to explore a CIM initiative.

CIM is easy to recognize when its technology is seen in operation, but it is difficult to initiate because it does not look like business as usual. The problems to be addressed at start-up are the lack of a management-level statement of the business concepts behind CIM, no defined management procedure on how to make it happen, and limited in-house expertise among the business groups which must cooperatively apply CIM.

The problems in going from installed examples of CIM to the pursuit of one's own program require tenacity. The first step is critical. And perhaps only those who can envision the full potential of CIM will press hard to overcome the start-up problems. Successful implementors have learned that it takes a deliberately organized and committed management team effort to go after CIM and to make it happen.

The key to moving toward a CIM plan is to face the critical management problems and issues. After all, CIM means developing a new way of doing business. To effectively launch this type of program, the management team must first prepare itself. The need is to develop perspective on the concepts inherent to a CIM business strategy and the practical considerations for a phased approach to applying such concepts.

Evolution of the CIM Business Strategy

The attractiveness of a CIM approach and its large potential benefits make it a timely strategic solution for industry's problems. In order to examine the specifics of a CIM business strategy thrust, it is helpful to review some origins of today's problems in the goods producing industry.

Discussions today are usually at a very high level and center on losing industrial competitiveness in global markets. The present situation evolved over many years. Included was a very simple tactic expertly applied by underdeveloped competitors. Their approach was to erode established industry leaders by penetrating a selected product category. Some examples of the product categories targeted were radios, TVs, cameras, etc. The success of these tactics is now history.

Competitor's criteria for the selection of a product category to penetrate were fairly standard. Most of these products had high demand, stable characteristics, firm markets, high volume, limited options, entrenched production processes, policies, and procedures.

Yesterday's Winning Manufacturing Tactics

Penetration tactics were applied to existing factory production facilities. The method was to examine how products were being made in minute detail. The objective was to pinpoint weaknesses across the many activities involved. Some examples of the most evident weaknesses were obsolete production machinery and high customer demand being allowed to override quality considerations. Escalating costs, such as inventory, were being absorbed by raising prices and not by increasing productivity.

The situation was aggravated because factory operations were not high on the interest lists of the management and technical communities. Many years of short-term paybacks and neglect of long-term investment were prevalent.

This condition made a once mighty production system easy prey for a new competitor's modernized upgrade. The competitive tactic was to apply a systems approach to restructure for strength by displacing inefficiencies wherever they existed.

This oversimplification of some thirty years of applying and refining this tactic for many different types of products is not an attempt to explain history. The point being emphasized is that the primary target was products in the factory. Secondly, there was no single weakness in the production of products which required new or advanced technology. The formula for competitiveness was to apply improvements in many parts of the production process. Yesterday's successes were mainly tactical—application improvements in the production of products.

Future: Strategies Win Over Tactics

To take an industrial leadership position today requires far more than repeating tactical production improvements and investments. The strategic composition of CIM is that it goes beyond past tactics of restructuring for strength with a factory focus. The future thrust is to move to a new plateau of industrial competitiveness across the entire product cycle. In this manner, CIM presents a new creative approach for manufactured goods and is not one of imitation with improvement.

To pinpoint the new business strategy CIM brings, some observations in this book will be made on the experience of installed users. The intent is to gain insight into what are common denominators of these unique installations. The generalizations derived are offered for broad consideration and application.

The CIM installations today exhibit two key characteristics which result in major gains. First, a product is selected as the focus for development of a CIM business strategy. Second, the product serves as the common vehicle to build new operational relations across the entire business product cycle. The generalization derived is that the CIM strategy for leadership is to operationally integrate design, manufacture, and marketing activities as these deal with the selected CIM product. Because of the lack of a term, this approach will be referred to as a business strategy for CIM-Product Operational Integration. A way of representing a CIM-Product Operational Strategy and elements is shown in *Figure 1*.

Referring to *Figure 1*, the elements numbered two through eight are the traditional functions required to bring products to market. Element number one is the new ingredient to be uniquely created to pursue a CIM-Product Business Strategy. It structures the new way information efficiently links the traditional functions, elements two through eight, across the entire product cycle. Element one results from a collaborative plan for handling information across the elements and organizations involved. The modern tool for implementing

7

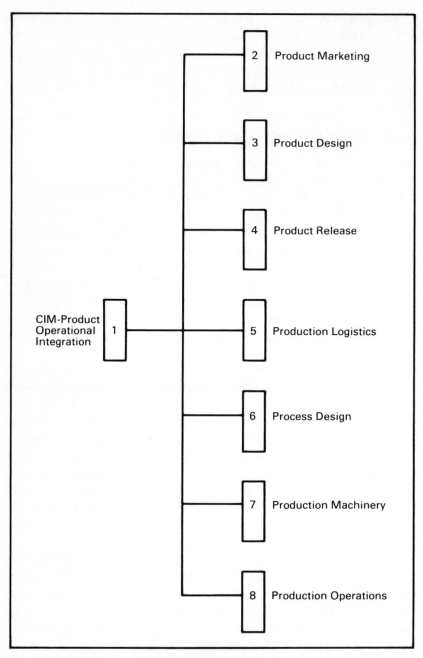

Figure 1
CIM-Product Business Strategy

this new operational integration is the information and communication technology of CIM.

The impact of a CIM-Product Business Strategy across the organizational functions is significant. An overview of this Enterprise Direction for CIM is summarized in *Figure 2*.

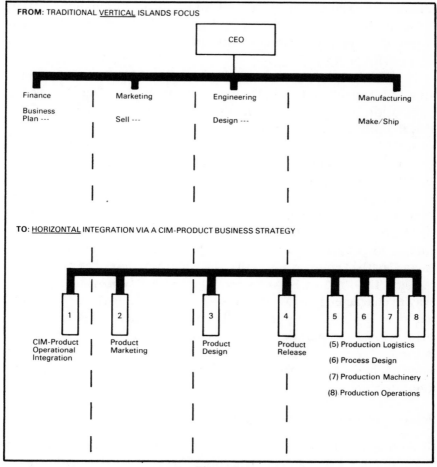

Figure 2
Enterprise Direction for CIM

CIM Strategy Priority Setting

The management of existing CIM installations generally agree on the long term goal for the integration of all the elements of CIM. However, each has set a different route and priorities to get there. The filling of their CIM strategies are uniquely determined by their product and business requirements.

Some generalizations can be made which relate the type of products involved and how these affect the priorities of integrating the various elements of the CIM strategy.

A firm that makes a product in the higher-price range which is engineered to customer order tends to fill its operational integration structure top-down, as shown in *Figure 1*. In contrast, a firm that produces a low-priced, standard product is likely to follow a bottom-up priority plan.

For a higher priced, critical replacement parts business, the emphasis is to fill-in integration of the center elements first. This stresses the integration of digitized product definition data as it relates to data-driven production machinery.

A CAD/CAM (Computer-Aided Design/Computer-Aided Manufacturing) type of product has program priorities for integrating strategy elements 3, 4, and 7 (see *Figure 1*).

For high demand, medium-priced products, the focus is on a bottom-up priority. This starts by concentrating on improvements to production operations, then adding more automatic machinery, then integration of production logistics. This pattern of events follows the tactical path of competition in the past which was described previously.

The CIM-Product Business Strategy (*Figure 1*) provides a general way to characterize CIM. It offers a way to position the CIM opportunity which is business and product niche oriented. The experienced leaders in CIM report that it requires four to six years to fill-in a major portion of their integration strategy. This level of investment deserves some guidelines concerning management's role in pursuing the CIM opportunity.

Management's Role in CIM

At this writing there is no general set of management procedures to address the CIM opportunity. This is because CIM is evolving by industrial practice and not by theory. A common message from successful CIM implementors is that management plays a key role from the very beginning.

In order to develop more specific information to help guide managers through the management cycle of CIM, some observations will be made on the experiences of installed users. The objective of this exercise is to sponsor more collective thinking on the management aspects of CIM.

To organize CIM into a sequential series of management initiated activities, the first requirement is to divide CIM into its phases of development. These represent the steps the management of a company would follow to determine its unique opportunities via CIM.

Based on observation, a CIM initiative may be viewed as consisting of four stages of development. Stage one is the initial INTRODUCTION to the new concepts within a CIM business strategy. The second stage is the PREPARATION necessary to develop background information to grow in general expertise on CIM. Stage three is the application of CIM to the particular situation at hand to develop a specific PROGRAM PLAN. The fourth stage is the IMPLEMENTATION of CIM. A summary of the types of activities occurring in these four stages is presented in *Table 2*.

TABLE 2
Management Cycle for CIM

Stage 1   *INTRODUCTION*
       (a)   Concepts of a CIM Business Strategy
       (b)   Managers' Roles
       (c)   Creating Interest in CIM
       (d)   Developing the Climate for a CIM-Initiative

Stage 2   *PREPARATION*
       (a)   Forming a Study Team for CIM
       (b)   CIM Opportunity Candidates
       (c)   Conceptualizing CIM Way of Operating

Stage 3   *PROGRAM PLAN*
       (a)   Proposal Development
       (b)   Selling Proposal
       (c)   Program Commitment

Stage 4   *IMPLEMENTATION*
       (a)   Managing the Implementation
       (b)   Measure and Evaluate Results
       (c)   Moving to Next Opportunity
       (d)   Sharing Experience and Expertise

Executing the CIM Cycle

The sections which follow have been organized to take the reader sequentially through the four stages of the Management Cycle for CIM. Each of the stages contains detailed descriptions of the activities required to go from the INTRODUCTION to CIM concepts through IMPLEMENTATION.

In this book, the authors share successful management experiences of their CIM teams—teams who have won the CASA/SME LEAD Awards for the years of 1982 through 1985. (LEAD is an acronym for Leadership and Excellence in the Application and Development of CIM.)

Future Direction

The future direction for industrial leadership continues to build on past and present progress. The industrial revolution brought new tools for progress to the factory. Mechanization by means of machinery gave muscle to build products. CIM now extends these capabilities while removing all previous bounds for systemization across product design, manufacture, and marketing. The newest tools for progress add informational intelligence to help manage the entire business cycle.

A CIM business strategy has been described as essential for leadership in the next plateau of excellence in the goods producing sector. This is a continuing evolution by extending the broad menu of computerized applications experience in industry. CIM is not to be perceived as starting some new type of project, but rather a single program direction to refocus existing and new projects.

Despite varied products and business requirements, a CIM direction reduces the business of manufactured goods to two elements. These are a CIM-Product Business Strategy and the new integrated operational relations this strategy development fosters across the entire product cycle.

The nature of a CIM program direction is to seek higher efficiencies and faster cycle times as a collective business unit across the entire cycle of product design, manufacture, and marketing. The management challenge becomes one to promote more collective participation by the many functional organizations affected. An effective vehicle is to develop a CIM-Product Business Strategy. This collaborative effort addresses CIM-size benefits and the efficiencies from sharing the combined assets of information networks.

CIM: MANAGING CHANGE FROM THE TOP DOWN

Kenneth W. Emery
General Electric Company

CIM's New Approach

The two aspects of CIM that must be kept clearly in mind in addressing management's role are those of computerization and integration. Ironically, in the past, the approaches of manufacturing organizations have been diametrically opposed to these concepts. The aim used to be to create structures which sub-divided work into elements small enough to be performed by an individual. With the tasks sub-divided, the next challenge was to create interface mechanisms to communicate the output of one individual's work to the next individual in the process and then finally to integrate all the activities through successively higher levels of management responsibility. While such an approach has worked well in the management of people, simply automating existing methods and procedures fails to account for the changes that computer technology introduces into the way tasks are performed as well as the impact the technology has on the existing work elements and environment. Simply automating existing practices does not result in CIM but in "islands of automation." Islands which in all probability are inefficient.

The first role of management, then, is to realize that CIM requires new ways of addressing both old and new problems. Management must also realize that CIM's new approaches may be neither fully understood nor at all welcomed by those in the organization accustomed to and comfortable with the old "standard operating procedures."

The Role of the CEO

In embarking upon CIM, therefore, the CEO must play a key role in terms of creating an atmosphere which does not accept the status quo. The CEO often is the only person in the organization with truly integrated perspective and responsibilities. If the CEO is not actively involved in the planning and execution of CIM, then the result inevitably will be on the automation of functional tasks without achieving the integration vital to the functioning of the business. The CEO's participation is three-part:

1. Planning the CIM effort so that strategic business needs are being addressed and to ensure cross-functional integration,
2. The building of a team approach toward CIM implementation and the development of enthusiasm, and
3. The enforcing of the long-range commitment to CIM for the business' good, even if it entails short-term difficulties.

Naturally, the CEO also has a role as the keeper of the purse strings. Typically CIM projects entail significant expenditures of money and manpower to be successful. The business must be willing to make and stand by these commitments to ensure that overall goals are met.

The Roles of Functional Managers

Functional managers play a variety of roles in the implementation and operation of CIM. They, like the CEO, must realize that CIM means new ways of doing things and that they have to react accordingly. They are team leaders for their functions in the CIM effort. Because the changes associated with CIM implementation may mean extra work with few immediate apparent benefits for an already overloaded organization, the functional manager must ensure that everyone understands the strategic importance of integration and the advantages it can bring to the total business.

The ultimate role of a functional manager is as the expert in the particular function. After all, the computer is just a machine, and the analysts and programmers are simply experts in its use. The real understanding of what is being automated and its relationship to other functions exists only at the functional level. Attempting to automate and integrate without this understanding will fail or, at best, produce little more than an automated documentation system which still relies on the people to accomplish all of the real work.

One of the major challenges in the CIM process is the reexamination of the functions an organization is doing and why the current practices work to the degree that they do. This process is a double-edged sword which, when applied conscientiously and with a totally open mind, will yield dramatic, unexpected benefits. Possibilities range from the total elimination of activities and documents once considered vital all the way to the addition of tasks which had previously been left to chance or to the judgement of unspecified individuals. Just as it is the function of the CEO to provide the climate for change, it is the role of the functional manager to implement those changes and to ensure both the added value and quality of the changes being implemented.

The CIM Planner

Should there be a focal CIM manager? Yes. In most organizations, the only manager whose perspective spans the total organization is the CEO. In order to focus the CIM plan to address the business' strategic needs, CIM planners must be able to step back from the traditional, functional relationships and view the business and their own responsibilities from a total perspective. This is the only way for the CIM planner to address the functions impartially. While it is theoretically possible to assign such a role to a person within an existing organization or to a task force or committee, the conflicting interests created by such an assignment will block the world view which is CIM's proper perspective. The second reason for identifying a focal CIM organization is to provide the opportunity to accumulate CIM expertise for subsequent, follow-on applications.

Unfortunately, today's work force is not as knowledgeable and comfortable with computers as the upcoming generation. As long as this situation exists, there will be a need to augment the skills of the general work force in this area. A CIM organization provides the opportunity to develop and exploit these skills. In the creation and implementation of these roles, the difference between CIM expertise and computer systems expertise must be clearly delineated, with computer systems expertise being only a small part of the total CIM picture.

CIM means doing both old and new things in new ways. The integration concept makes CIM a total business activity. Every component in the organization has the charge of bringing about this change and taking ownership for the integration.

OVERVIEW OF MANAGEMENT'S ROLE IN CIM

Paul W. Brauninger
Cone Drive

The key to bringing the concept of Computer-Integrated Manufacturing into our everyday business practices lies within management initiative. Management, and specifically the chief executive of the organization must provide the stimulus to generate the interest and motivation to institute a CIM program.

The Role of the CEO

Each enterprise has its own unique personality and social structure, but the chief executive must be the one that guides the business toward a common goal of CIM. The chief executive must clearly communicate to subordinates the CIM objective and ensure that continual motivation is applied in all areas. The chief executive's primary role is in building a team approach and setting the proper environment for a CIM program.

The chief executive should understand that minor setbacks may be encountered when undertaking new ventures or risks. Because of this, the chief executive must encourage his or her staff to persevere and achieve the ultimate objective...profit through Computer-Integrated Manufacturing.

Cone Drive believes that people are the key ingredient for a successful CIM program. It is essential that the chief executive ensure that the staff also believes in CIM. Without a staff that is totally committed to the concept of CIM, the successful implementation and the whole program is in jeopardy.

The Role of the Functional Manager

The functional manager is the next layer in the organization that must also believe in CIM, and openly support the program. This level of management is the most critical element to the overall successfulness of implementation and the overall program. If the functional management level is not solidly committed to the program, their subordinates will sense this, and they in turn will be against the program.

The manager must also assign the most capable people to the project. Ideally the person will be someone respected by the rest of the group, have leadership skills, and be an experienced enough individual to understand how things are accomplished in this particular enterprise. The person responsible must be someone who has an understanding that schedules must be met. Knowing what is required and accepting responsibility for timely completion is essential. An individual who accepts or makes excuses for why things weren't done is totally unacceptable.

Since most projects require the interaction of different functional areas of the organization, it is also important that no one group is dominant in forcing its ways on other functional areas. A team approach is necessary to achieve optimum results. However, the team does require an individual who can provide direction to ensure that all the needs of the organization are adequately addressed and to ensure that decisions are made in a timely manner.

The CIM Project Leader

The key ingredient in managing the CIM project is the project leader. The chief executive must ensure that there is a team approach but must also ensure that the responsibility for getting things done is given to one individual. This individual will be the one that oversees the project, ensures that each group is meeting its particular milestones, and advises the chief executive on the progress, lack of progress, or problems requiring attention.

The Importance of Motivation

Having highlighted the importance of top management properly setting the right climate and motivation for CIM, management must also ensure that the individual managers and supervisors also have a positive attitude toward CIM. Cone Drive has discovered that the most difficult individuals to motivate in a CIM program are the middle managers and the shop foreman. There were a variety of reasons for this attitude. However, typically the reasons related to past experiences with projects being initiated with a lot of fanfare only to have another project come along, allowing the previous project to be forgotten.

In other cases the resistance was caused by a natural resistance to change. Even though the existing system may have caused individual managers some problems, they knew how it worked. Changing their routine or initiating new ideas was not readily accepted. While initiating a new idea may have benefited them, if they did not immediately understand the project or idea, they felt their leadership role was being challenged or their job positions were in jeopardy. Obviously, this type of fear will hinder the acceptance of any new system. Therefore, it is essential that upper management understand the needs and fears of the overall organization prior to implementation.

Stage 1
Introduction to CIM

Stage 2
Preparation for CIM

Stage 3
Program Plan for CIM

Stage 4
Implementation of CIM

SPARKING INITIAL INTEREST IN CIM

Paul W. Brauninger
Cone Drive

A question asked of Cone Drive on numerous occasions is: How did we initially become interested in Computer-Integrated Manufacturing (CIM)? Initially, Cone Drive was not out to become a leader in CIM technology, or even to implement a CIM program. What started our interest was customer demand and competitive pressures. We realized we had problems with our delivery integrity and our product lead times.

Competitive Pressures Drive CIM

Originally, we had attempted to solve this problem over the years by hiring additional people, building more inventory, and extending lead times, but achieved limited if not poor results. The lack of success was very frustrating to the entire organization. Competitive pressures forced us to look to alternatives that might achieve better delivery integrity along with shorter lead times. We realized that we had to streamline our organization and update our equipment to satisfy our customers' requirements if we were to ensure the survival of our business.

It became evident that to ensure our survival we had to do things better than our competitors. This forced us to look at a number of alternatives. We chose an alternative that involved the use of computers and other related technologies involving systems such as MRP (Manufacturing Resource Planning) and CAD/CAM (Computer-Aided Design/Computer-Aided Manufacturing).

Top management realized that something had to be done, and provided the personnel and financial resources to get started. Two fundamentally important ideas were developed initially... First, "find something that works, and use it's capabilities to the fullest instead of forever looking for the ultimate solution"...Second, "make sure every department is involved from the beginning." These two ideas were very important in our CIM program because they caused us to develop a

CIM plan for coordinating the efforts of each department. These two ideas also helped to generate enthusiasm and a sense of teamwork within the organization.

"It Won't Work for Us"

We started our CIM program by addressing the most serious issue to our business survival, which was inventory control and delivery integrity. Immediately, we heard from our managers "We're unique" and "Things that work for other people won't work for us." They believed that our organization was so special in nature because of past practices that it was not feasible to make effective changes in the organization.

To overcome this idea, we organized visits to other companies utilizing computers, and attended seminars, shows and conferences on CIM, MRP, CAD/CAM, JIT (Just-In-Time), and automation. Through this exposure, our people came back with the idea that we were not that unique. Other companies had said the exact same things; but they had actually implemented these new technologies, and the new technologies worked effectively. A wide variety of management levels were exposed to these presentations, including top management, middle management, and individual employees to ensure that everyone was involved and informed.

The visits, seminars, and conferences also achieved an unexpected result. It made our people realize that there were different ways of accomplishing tasks. It initiated thoughts about possible changes in the ways that we accomplished things. This was probably the single most important event in initiating interest in CIM...the idea that there may be a better way.

Initial Success

Looking back at our CIM implementation program, probably the single most important event in maintaining interest in CIM was the success we achieved with our first system. We were able to demonstrate to ourselves that CIM could work for a company with our product characteristics and size. Even the organizational "doubters" that went along with the program but said "it won't work here" changed their attitudes.

Our success with the first CIM project certainly helped in the financial justification for our future projects. The degree of risk associated with venturing into another CIM project was greatly reduced after our first success.

Getting Everyone Involved

In summary, the primary method to spark interest in CIM is to get everyone involved. The following list includes what we at Cone Drive used to spark interest in CIM:

1. Developed expectations that things were going to get better— that new systems were going to be better than the old ways of doing things. This got all employees a little excited. They knew that the way things were done was not the best way, but had never been given the tools to do a better job.
2. Trained employees in the new system before the actual installation. This developed confidence that they could do their jobs using more advanced systems and equipment. We also promised people they would be trained in a new (and probably better) job if their current job was eliminated. This gave them security and a reason to look for time savings.
3. Remained open to suggestions for improvement whether it related to specific tasks or the system as a whole. Although the basic jobs needed to be done, we were able to modify many tasks to be done the way the operator wanted.
4. Allowed people with ambition and desire to lead in the implementation of new ideas and projects.
5. Showed everyone the results as projects were successfully completed.

The enthusiasm generated by our first CIM project has since been enhanced by the successful implementation of our CAD/CAM system. Now our challenge is to manage our future programs so that we do not attempt to accomplish too much at one time. Our past experiences have allowed us to establish a consistent tempo for our future implementation projects to fully integrate our current operation.

EXECUTIVE PERSPECTIVE ON CIM

Wade L. Ogburn
AT&T Technologies

Educational Briefings

The football cliche, "The best defense is a good offense" is appropriate when describing how to spark the initial interest of upper management in Computer-Integrated Manufacturing (CIM). Upper management is often quick to say "no" to implementing CIM when confronted with the large outlay of capital without prior experiences. Their interest should be sparked via a planned series of educational briefings on CIM. It is imperative that management learn about the many opportunities of all affected areas of the company. A well prepared CIM offense is vital from the beginning.

Remember, keep the first introduction concise and simple. Do not overload executives with too many facts. This will inhibit instead of spark their interest. An off-site location would be advisable. This would prevent the normal business day interruptions that would occur if the meeting were conducted on-site. Encourage management to investigate the efforts of other companies, professional societies, and universities. If enough interest is generated to cause management to learn more about CIM and start providing their input, the chance for a "yes" decision is greatly enhanced.

The benefit areas of CIM should be investigated. Every member of upper management has their own particular area of interest. This area should be addressed individually and collectively. Explore the CIM advantage with the executive. What are some of the potential benefits of CIM? There are many and this article cannot cover all of them. A few will be discussed to enable one to get started:

1. Increased Productivity—Companies usually experience manufacturing intervals shortened by 50 percent or more. Computer-aided manufacturing (CAM) shortens the elapsed time for the manufacturing cycle. Numerically controlled (NC) tools and Flexible Manufacturing Systems (FMS) give the capability to have unmanned operations. Industrial robots have experienced

productivity gains of 300-400%.

2. Engineering and Design Cost—Computer-aided design (CAD) reduces lead time and cost of introducing new products into production. Also, changes to existing products are easier to control and introduce into production without disrupting the manufacturing process. Group technology (GT) allows for standardization of routing and computer control files, eliminating manual entry of data. This reduces errors, redundancy, and lead time.

3. Product Quality—Computer-aided manufacturing (CAM) makes the work force more reliable by taking the guesswork out of manufacturing. Industrial robots relieve people of monotonous and often dangerous jobs. Numerically controlled machines increase quality because less skill is required by the operator. CIM will reduce scrap and rework, especially in an environment where one has to adhere to strict tolerances. When there is scrap or rework, CIM gives one the ability to discover the reason for the problem and allows one to take remedial action more quickly and with a greater degree of success.

4. Production Control—A Computer-aided manufacturing system will have an integrated database. All data needed by production control to interface with manufacturing, engineering, purchasing, and the customer is readily accessible. Production schedules are computerized, allowing the computer to become the expediter. This frees the production analyst to solve problems that were ignored in the pre-CIM era. Also, with CAM, less paperwork is required and this increases the efficiency of all parties involved.

5. Capacity Utilization—An integrated base provides the capability of performing facility studies helping in development of capacity plans. Also, "what if" capability prevents surprises in the future. Numerically controlled tools have less downtime due to off-line programming; and they require less floor space.

6. Customer Satisfaction—All companies thrive on repeat business. If the customers are satisfied, they will continue to give your company their business. Along with all the advantages already listed, CIM will give your company the capability to provide shorter service intervals at lower cost. CIM's flexibility will provide a better planning tool and the capability of being

more responsive to the customer. The simple capability of giving the customer an accurate answer to questions is great for customer relations.

7. Social Factors—Some benefits of CIM are hard to measure. One of these is the social benefit. Employees will have a higher morale because of more challenging jobs, less overtime, and in some cases, less dangerous jobs. Also, the additional training usually leads to an upgrade, builds self-esteem, and results in the employee becoming a contributing member of the overall process. An improved working environment is one of the many intangible benefits of CIM.

GETTING THE CEO ON BOARD AT THE START

John F. Snyder
General Electric Company

The heart of CIM is multi-functional integration. It's neither an MRP II (Manufacturing Resource Planning) system, nor a flexible manufacturing cell, nor interactive graphics, nor office automation, nor computerized market forecasting, nor the latest general ledger system. These are simply "islands of automation"—powerful but lacking the synergy that comes from digitally linking them.

Because CIM cuts across functional lines, top management commitment is needed to drive successful implementation. If ownership of CIM is viewed as belonging primarily to a specific function, less cooperation will result. If CIM is viewed as a system project rather than a strategic thrust, synergy is threatened. And if for no other reason than that CIM implementation can require committing substantial, long-term resources, the CEO needs to be the ultimate leader and driver of the CIM effort. Major programs expected to make quantum changes in a business' productivity and competitiveness cannot be expected to start at the bottom of a hierarchy and trickle up.

Getting the CEO on board from the start involves not simply educating in the technology but also driving home the linkage between CIM and the business' needs in a competitive market environment. What follows is simply a variety of techniques to raise the CEO's initial interest in CIM.

Approach the CIM Issue Strategically

Data communications among computer systems, for one example, is an exciting, challenging facet of CIM. It can also be a deadly boring technical detail for someone whose sight is focused on the bigger picture. To interest the CEO, start with CIM and business strategy. Here's a sample of typical strategic questions that can point to a CIM program. Many more can be identified.

What's the Competition Up To?

It often is fairly easy to get a reading on a competitor's level of systems and automation. This can be gleaned from publicly available information such as articles and references in trade and professional journals. Brochures are easy to come by, especially at trade shows, and often provide insights into manufacturing, assembly, and design processes and sophistication. Customers will often freely share what they've learned from the competition during the marketing effort, and many times a vendor will have a handle on the competition's automation efforts; the CEO is usually quite interested in catching up or pulling further ahead in order to ensure the basic health of the business. Enough results are in on CIM productivity investments to make a strong case that a business is under serious threat if it falls behind in the CIM effort.

What's Important to the Customer?

Price, sure, and CIM can make dents in costs, but what about consistent quality? Even when the product can be considered a commodity, CIM can address differentiation through quality. When interfaces between functions are mechanized under CIM, a consistency develops that eliminates the mistakes that occur when information and data are passed manually, verbally, or in incompatible formats that have to be reinterpreted. And how about delivery in time? Is that important to the customer? Most manufacturing operations in the

U.S. deal with small batches of a sizeable variety of products. CIM's automation of the interfaces not only between functions but between activities within functions realizes significant reduction in cycle times.

How Has Overhead Been Behaving Recently?

The media these days focuses on what they can take pictures of, giving the impression that automation is equal to robots which are equal to direct labor reductions. Actually, the white collar worker, depending on meetings and paper to get the information needed to do the job, is a prime target of CIM productivity gains. Again, unless education takes place, CIM gets equated with the "islands of automation," overlooking the mammoth manual efforts and paperwork often needed to keep these islands afloat. Even a cursory audit of the manual effort that goes on between existing systems can be an enlightening—if not frightening—experience.

Expose the CEO to Those Who Speak the Same Language

There are enough businesses around with successful CIM implementation stories for one to be able to arrange a visit to see CIM in operation. Such sites usually are happy to showcase what has taken prodigious effort to pull off... and their enthusiasm can be infectious. It's important to get that kind of exposure for the CEO. But it's not enough. After a few site visits, set up a meeting or a luncheon date for the CEO with the CEO's counterpart of one of the sites visited (typical tours generally tend to trot out technical leaders rather than top management). Have it be one-on-one with no CIM missionaries from either side attending so that the CEOs can be candid.

Educate the CEO to CIM's Impact

Once the CEO has had a taste of the strategic importance of CIM and had a flavor of CIM in action, it's time to pull together an overview of how CIM would apply specifically to the CEO's business, with both its risks and rewards. This is a simplified "master plan" comparing in broad terms how the business operates now as to how it would look with CIM; rough projections of resources that would be required for no more than three scenarios of varying CIM sophistication; projected benefits of each scenario based on general industry experience; a hypothetical time-table for implementation; and a discussion of the risks involved of both implementing and not implementing CIM. If the CEO buys into CIM, then is the time for the rigorous planning effort to flesh out the full impact and needs of CIM and to set in place the structure, reporting mechanisms, and standards for implementation.

SETTING A CLIMATE THROUGH STRUCTURE AND PLANNING

Kenneth W. Emery
General Electric Company

The first step in creating a climate to implement CIM is to understand the existing organizational climate: why it is the way it is and how it might be changed. Before the advent of computer systems, organizations were created primarily as a method of subdividing tasks into elements small enough to be completed by individuals. Interfaces for each subdivision had to be created to facilitate the flow of information among individuals and organizations. These interfaces consisted mainly of forms or other paper documents. Through time, bureaucratic forces institutionalized much of this interface documentation, and their production was looked on as the primary function of the individual or organization.

Computers were applied at first primarily to perform the massive computations associated with the financial function. The result was that systems programming and the first mainframe computers came under the ownership of the financial function. To computerize, armies of analysts and programmers were hired within the financial function. As manual activities were automated, other functions saw the future systems.

With a clear vision of what is versus what will be, the next step is to identify the steps that lead from the present into the future. Each program can then be evaluated in relationship to the overall plan. Resource requirements can be identified, and the business can begin to move in a single unified manner toward the common objectives.

Within such a framework, systems development can occur in "bite-size chunks" which are practical in terms of meaningful short-term efforts and benefits. Initial systems development can form a base upon which to build toward the objective of the long-range plan.

CIM activity may well start with a prototype system to test out the concepts, identifying the unanticipated and proving out benefits. The use of a prototype system helps set a climate by providing a showcase. A prototype also can help show that change is being managed realistically. There is a great deal of comfort in knowing that when a new system is introduced in one's section, it's been proven out beforehand.

The final climate setting activity is establishing formal review mechanisms for the CIM master plan in order to chart progress and to take into account changing technologies and business needs.

FORMING A TEAM TO SEEK CIM OPPORTUNITIES

John W. Pearson
AT&T Technologies

When the opportunity is presented for developing a plan to introduce CIM concepts into a business, it is of considerable value to form an action team. The team should begin preparing a business plan to realize the benefits of such technology. The team approach provides certain benefits to long-range business planning that are superior to individual planning. The synergy of a team working toward such a goal can provide considerably more effective integration of the business operations than one or two individuals that represent a narrower section of the organization.

The effectiveness of using a team in creating CIM development plans can be significantly affected by a business' position in its evolution. In the case of a new manufacturing facility, the appropriate time for use of a CIM planning team is immediately after the decision has been made to implement the factory. This was precisely the situation that led to extensive use of CIM concepts in development of the Richmond AT&T manufacturing plant.

In another prevalent situation, that of adding CIM technology to an already existing facility, such a team should be formed when the financial benefits can be realized and when management is deemed to be receptive to enhancing the manner of operating the business. The inertia of an already productive business may, under some circumstances, pose additional constraints to introduction of CIM technology if existing personnel do not feel that such changes are in their best interests. Because of this, there may be more "selling" needed to convince all persons affected that changing some aspects of the business will be in the best interest of all.

Forming the Team

When the climate is right, an effort should be put forth to generally define what goals and benefits are expected by introducing CIM. When some idea of what is desired has been formulated, a multi-disciplined team should then be formed to further define the goals and generalize on how the goals can be accomplished.

In forming the team, members with a variety of training and experience should be included. Using a single discipline can seriously limit the effectiveness of the team in developing plans that will address all aspects of operating the business. The actual number of participants will vary with business complexity and size and business objectives. It would not be unreasonable for a team to consist of two to five persons in a small company to a dozen or more in a substantial business.

Managing the Team for Effectiveness

As with any significant business resources, a CIM team will require guidance and management support to effect a change in the method of operating the business and gain the full benefit of CIM technology. Given a reasonably free hand at designing the manner in which a business will be operated will allow the team to fully incorporate CIM into the fabric of the business. Any other approach would generally appear to be like placing Band-Aids on problem areas in the business without fully addressing cost and operational problems.

Persons charged with assembling and managing the team must plan for the team members to spend an appropriate amount of time on team activities to make progress in keeping with the timetable needed for the business. For planning a new plant, it would be appropriate to expect the team to meet from two to three half-days a week to full-time in planning activities depending on the timetable of the new facility. Planning a new plant that must come on-line in a year or two will require heavy effort initially and will likely turn into a full-time job for team members. Adding CIM technology to an already existing facility may require a limited amount of time and could stretch over a few years.

Obviously, top management should interact with the team and should stay close to what is being considered and planned. Guidance should be provided frequently as to the advisability of plans the team develops. Various groups that will be expected to use CIM concepts should be active to some degree in the planning stage so they will be more effective in implementing the plan.

Makeup of the Team

Generally, the team should include persons that have knowledge of manufacturing operations, production control, facilities engineering, process engineering, factory planning, and information systems development. For complete integration of the CIM functions, the team should also include members from the computer-aided design and production engineering segments of the business. Other support organizations may be tapped as needed for additional contributions. In a small business, the team members may have to bear more of the load and the team may not have the resources for a full team.

Global Objectives of the CIM Effort

If a team is formed in the early stages of planning a facility, one of the initial objectives should be to closely study the manufacturing objectives and existing methods to determine if and how those may be significantly improved with the introduction of CIM concepts. It is not unreasonable to recommend the exclusion of CIM from some facilities since the cost of such technology may actually outweigh the benefits if the complexities of the products or the volume of production does not warrant the use of such methods.

After a determination has been made on potential gains to be made by incorporating CIM, a review should be made with management personnel to present the findings and obtain continued support for the team goals. This is a dual purpose effort, in that management must see a definite cost effectiveness from the investment into CIM technology. Supportive management is the key to continuing the existence of the CIM team and expansion of the effort to implement CIM into operations of the business. Appropriate course adjustments should be made to include any new changes in the expectations and needs of the business, and to tune the objectives of the team.

A considerable amount of time could be consumed by the team reviewing the needs of the business. Therefore persons involved should be familiar with the business upon joining the team. Efforts should be directed toward reviewing current CIM technology and determining how well it can be adapted to the business at hand. Use of a multi-disciplined team can avoid many pitfalls where the business objectives or technology to be dealt with exceed the experience or training of inexperienced team members.

The ultimate goals of a "CIM team" might generally follow (but not be restricted to) these items:

1. Form team, study needs of business, and determine if CIM is appropriate for objectives of the business.
2. Review market for range and sources of technology that may be appropriate for business goals.
3. Formulate plan of how facility will be developed or changed using CIM as integral part of the business.
4. Participate in actual planning of facility and development of plant and equipment, as well as structuring operating methods to be used in running the business.
5. Frequently, team members will have an opportunity to continue participating in CIM implementation after development of the long-range plan. Opportunities may be available to help build and start the facility, and to ultimately join the staff of personnel operating the facility.

In a large business, persons contributing to team efforts such as these may possibly be involved in a continuing effort to expand the use of CIM into other company locations and operations. Obviously in a small organization the efforts may be limited to a single or occasional involvement in CIM implementation. The individuals may have normal support responsibilities during the time of CIM planning or may return to normal activities soon after the team goals have been met.

Vendor Use/Participation

CIM team members should not overlook the obvious advantages of using vendor resources in planning for CIM implementation. Vendor resources should not be allowed to perform the actual business planning since a vendor would not necessarily have the best interest of the business in mind. On the other side of the coin, vendor knowledge about available technology and integration techniques can be quite valuable in the planning effort.

The team can tap vendor resources at various stages of the planning and implementation process to improve the fit of the CIM plan to the business. Vendor information is necessary; it helps in choosing various options in equipment, methods, and services during the planning effort. Additional information about costs and alternatives will be necessary after a plan is finalized and a commitment is made to implement the technology. Vendor participation will be necessary during stages where bidding is used to determine final costs and select suppliers.

A Funny Thing Happened on the Way to Developing the CIM Team

Early involvement of a team representing various disciplines needed to plan, develop, and operate a business will considerably increase the chances of successful implementation of CIM technology. Continued involvement and support of management is crucial to the team activities, success, and future. Keeping management informed and involved will go a long way toward ensuring timely implementation and financial success of the effort. The CIM team must be part of and participate in the larger role of developing and operating the facility in order to realize the full benefits of CIM technology.

A business can gain more than just implementation of new concepts such as CIM by forming an implementation team. Occasional review of the current or expected operations of the business and how it may be improved can be of distinct advantage to the business. Current operations could be significantly improved and cost of operations may be reduced through addition of such technology. Use of the team will increase the chances of thorough incorporation of the technology into the business and reduce the risk of developing pockets of automation not integrated into a total system. Even if the decision to not heavily integrate were made, the team can generally find areas to effect cost savings and improve manufacturing responsiveness through involvement and participation in the process of looking closely at the operations of the business.

CIM TEAM IS OUR MANAGEMENT TEAM

Paul W. Brauninger
Cone Drive

Teamwork

At Cone Drive we found that one of the fundamentals in establishing a CIM program was to develop a sense of teamwork. The team approach was critical in establishing a climate where the success or failure was a group effort and no one individually could succeed or fail because of the program. We had found in the past that sometimes programs were looked at by others as special "pet projects" of certain people or departments. In addition, since a CIM program was actually a new way of doing business, (indeed, the future of our business) we needed the input and resources of every department. And finally, we needed a method to visibly show to everyone that this was a critical program and that it had the support of top management. What better way to get top management involved than to make them a part of a new thing called the CIM Team.

The decision as to who should be members of the team was very easy because our objective was to have all the different functions of the organization represented. To establish a team effort, each of the functional managers reporting to the General Manager along with the General Manager made up the team. The membership consisted of Engineering, Manufacturing, Materials, and Data Processing Managers, along with the Controller and General Manager. In addition, certain other functional managers rotated in and out of the team, such as Sales, Quality Control, and Personnel. With this level of representation, and because every function of the business was represented, we felt confident that we could identify what needed to be done and that we could allocate the resources required to complete the task.

Objectives

The major objectives of the CIM team were to visibly show Management's commitment and to establish a mechanism where multi-functional decisions could be addressed and then timely decisions made and communicated to all employees. We found that in many cases, the ability to make timely decisions was a key factor in our successes. We had some instances where we had to go back and change a decision that we had previously made, but by making the decision we were able to do other tasks that we would have been prohibited from doing until a decision was made. The quick decision ability also helped to show the urgency of the program, and management's personal involvement in its implementation.

The decision as to who would chair the committee was also simple. We used the same chain of command as was normal for our business. The General Manager was the chairman of our team. We were very fortunate in having the level of involvement we had from our General Manager. This involvement was critical in communicating to everyone the importance of the program to the long-term future of the business. Also, by having the General Manager head-up the team, the instances of internal bickering and provincialism were kept to a minimum. A critical element in having the General Manager as part of the team ensured us that when a decision was made, it had the full backing of everyone involved, and eliminated second guessing.

Team Operations

The first function of the team was to set up a regular meeting schedule for the future; we started with a bi-weekly schedule. The next item was establishing basic criteria for new systems such as using canned software versus writing our own. Where do we start shop floor versus office? Should we use our people or hire consultants? And, what about a general timetable for getting things done?

As mentioned previously, the team started meeting bi-weekly, but whenever a critical decision needed to be made or direction redefined, the team met to review the situation and make a decision. In some cases, the team met weekly. The attendance at the meeting varied a great deal; we tried to have people from departments or functions that would be affected express their opinions and input as to what direction we should take. We also included outside vendors and experienced users of systems in our meetings whenever appropriate.

In summary, the CIM team was our management team. By using this existing team, decisions were made and carried out very efficiently. It also achieved the sense of teamwork and top management commitment so essential to the success of any project.

CONCEPTUALIZE THE NEW WAY USING CIM

Donald B. Ewaldz and George J. Hess
The Ingersoll Milling Machine Company

Conceptual Specification

It is very hard for each functional manager to visualize the "new way" (of doing business) with CIM from a specialized vantage point. It is very much like the three blind men feeling different parts of the elephant and trying to give their impression of the whole. The big difference is that the three blind men can do no harm to the elephant because he has already been created, but if the functional managers attempt to construct an integrated system from diverse visions of what the finished product will be, disaster is assured.

How do you prevent such a disaster? You prevent it by preparing a "conceptual specification" of what the "new way" of doing business will be. This gives all of the people in all of the functions a clear and unified "concept" of what the overall new way of doing business will be.

Strategic Planning

However, before serious planning can take place for the CIM system, the overall strategic plan for the business must be developed. This will describe the "general nature and intent" of the business, the "served markets", the products that are in production today, and those that are in the product planning cycle. It will describe the organization and facilities needed to excel in the production, installation, and field support for these products, leading to satisfied and repeat customers. With this as a background, then serious planning of the CIM system can proceed.

Creating the Specification

Construction of the conceptual specification is an iterative process. There is no cookbook approach to guide any particular organization to the perfect solution. One starts with a small team, or even a single architect outlining his or her vision of what the new way of doing business "could be". The selection of this architect, or team of architects, is probably the single most important key to success. The team must have a delicate balance of the practical needs of today's business, and a truly visionary understanding of what it could be if all available known and emerging technologies were exploited to the fullest. This vision of what the new way of doing business could be is then clearly communicated to all of the functional managers of the business (marketing, engineering, manufacturing, and accounting). This must be carefully communicated to them in a non-threatening way to develop in them a constructive, supportive, and creative frame of mind.

Each of the functional managers working with his staff should then modify this "could be" draft into what it "should be" to best fit the individual function within the future vision of the business described.

The functional "should be" drafts are then returned to the coordinating architectural team to be integrated together into a final "to be" conceptual specification. Just as the name implies, such a document describes the concept of what the new way of doing business is "to be".

The above description, although technically correct, is deceptively oversimplified. It describes the process as iterative, but then implies that it is complete in just three iterations. This, of course, is seldom if ever the case. If the architectural team is a strong visionary team as it should be, and the functional managers really get involved as they must be to make this a success, there will be numerous iterations of the "could be/should be" step. However, when it ends, there is just one "to be" final document understood and enthusiastically endorsed by all.

Feasibility Study

This conceptual specification is what is used as a basis of the feasibility study to evaluate the cost savings or benefits to be derived from the proposed implementation of this new way of doing business with a computer-integrated manufacturing system. The feasibility study, with the conceptual specification as supporting documentation, is submitted with a specific "request for expenditure" for approval of the project.

Stage 1
Introduction to CIM

Stage 2
Preparation for CIM

Stage 3
Program Plan for CIM

Stage 4
Implementation of CIM

STRUCTURING A CIM PROPOSAL

Donald B. Ewaldz and George J. Hess
The Ingersoll Milling Machine Company

Preparing the Firm for the Proposal

Bringing a firm from an environment in which computer technology plays only a limited role—or no role at all—to one in which all the functions of the firm are integrated through computer technology is a major and sometimes traumatic undertaking. Capturing the benefits computer technology offers may require drastic changes in the organization of the firm; it will certainly mean drastic changes in the way information is distributed and received. Traditional departmental boundaries will be challenged. Information once closely held and meted out by a particular function will now be immediately available to any who needs it. There will most certainly be staunch resistance to computer integration, if the way is not carefully prepared. There is often a strong tendency to continue traditional manual information systems, bypassing the computer-based alternatives, if the users aren't comfortable with them—or don't feel "ownership" of them.

Consequently, the real proposal preparation work must begin long before the preparation of the actual proposal document, and must begin by preparing people in the firm culturally and psychologically for computer technology.

This won't be easy; traditional barriers are often difficult to overcome. However, they must be overcome, if the investment in computer technology is to succeed.

The first step, in proposing computer integration, is to informally and diplomatically introduce the concept; to persuade peer and senior management of its benefits and show them that the value matches the effort and change needed. (That doesn't mean one must attempt to train senior managers in computer technology. Trying to make programmers out of executives is a mistake, and sometimes a tragic one; the real objective should be to show the potential advantages in terms of streamlining the operations of the firm, in reducing operating costs, and in improving the firm's responsiveness.)

The concept of computer integration must have the enthusiastic commitment of the senior managers of the firm; anything less will compromise not only the success of the proposal, it will severely jeopardize the success of carrying out the activities of integration even if the proposal is accepted. Necessary changes are going to face resistance. Major investment commitments are going to be needed. Some elements may fail, and must not be allowed to destroy the credibility of the overall concept. The only way these barriers and risks can be accommodated is by making the proposal a true joint effort. If one develops a proposal for computer technology just as one would go about developing a proposal for a machine tool or building, one fails to understand the sweeping implications of the opportunities computer technology offers.

Forming the Proposal Team

The impact of computer technology will be felt throughout the firm. Consequently, the formal proposal preparation must be a joint effort of the managers of each function of the firm. There is often a question of what functions should be included in the team; the best answer is that each operation function must be included. Most firms recognize the need for including the accounting, design engineering, manufacturing engineering, and production control functions; but some fail to see the critical need for others, such as marketing, human resources, and purchasing. The need is often recognized too late.

The proposal must be the product of the managers of the functions. The most significant benefits of computer integration stem not from the traditional target of cost reduction in factory direct labor. The traditional "overhead accounts" often contribute as much as 50% of the total manufactured cost of goods, and offer the most dramatic cost reduction opportunities. The functional managers must be committed to capturing such opportunities, to making reductions in staffing levels, to changing operation policies.

Materials and marketing management must be committed to slashing inventory because they recognize and believe in the streamlining that computer integration can contribute. That can only happen if managers, not their representatives, are an integral part of preparing the proposal. It often makes sense to have the manager of data processing or computer systems lead the team, but the manager must be sure to only "lead"; a dictatorial team leader can destroy a CIM system by imposing ideas on the operating functions.

Clearly, this further dramatizes the need for doing an exemplary job in introducing the concept. Developing the proposal will require a sizeable effort on the part of each of the functional managers. If they are not enthused and committed to the task, the team will quickly wither, and the advent of computer integration will be delayed until someone does better in persuading them to the task. Senior managers must enthusiastically support the effort, not only in recognizing the time the functional managers are taking from their usual assignments, but to reinspire those functional managers who begin to lose interest.

Justifying Computer Technology

The benefits of computer technology are real, quantifiable, and easy to identify. However, the sources of the benefits are not the traditional target costs of industrial engineers and "time study experts". There are several major cost areas that are dramatically impacted by the introduction of computer technology."

All the costs that make up the overhead accounts are affected by CIM. For example, often a considerable number of clerical personnel can be eliminated, because filing, sorting, making record copies, the traditional major clerical tasks, can be eliminated. Information is available "on-line", and is self-filing. (Some firms have reduced clerical personnel as much as 70%.) The number of people needed for production scheduling can be cut dramatically, in most cases, because many of the scheduling tasks traditionally done manually are done by the computer.

Inventory levels can be slashed. The overall production system will be much more responsive to incoming orders, and heavy inventory won't be required to meet service levels. For example, one of the traditional contributors to long lead times is order entry; preparing the paperwork just to get an order "into the system." An on-line, menu-driven order entry system can have an order into the system and all functions aware of it in seconds. Orders can be placed to suppliers in small fractions of the time required in a paper-based system; some firms have established networks into their suppliers' master schedules, and orders are placed and scheduled electronically. This means there is a dramatic change in the way the firm does business. The net effect is reduced working capital requirements for the firm, and reduced interest cost for the investment, so both income statement and balance sheet are affected.

Overall productivity of the assets employed can be leveraged up significantly because each can be used more effectively and efficiently.

For example, engineering changes can be applied more cost effectively because the status of material in production is really known, and changes which cost more in scrap than they offer in value can be delayed. Schedules can be updated instantaneously, so all individuals are aware of current priorities, and are working toward the same schedule target.

These benefits all add up to big savings—savings in operating costs, in working capital levels, and in return on assets employed. They can be identified, sized, and their impact reliably predicted. Their potential value can justify major investment in computer hardware and software; one must only have imagination enough to recognize the opportunities, and to look beyond traditional justification procedures to the impact of the investment on the overall cost of operating the business.

Elements of the Proposal

The format for the proposal will vary, depending on the individual firm, of course, but there are some elements that should be "givens" for any CIM proposal:

The proposal must start with a well-written executive summary. This is essential to win the real commitment of the top executives. No one can fully support something with only a vague idea of what it is all about. Executives must have a crisp, clear picture of the computer integration concept; this may be the last chance the proposer has to win over top executives. It may also be all they have time to read. It had better be good.

The proposal should clearly define the overall structure of the computer integration package. It should identify the various modules—both hardware and software—that will comprise the system, and how the modules interface with each other. This description should act as a control, and should be in enough detail so that any of the modules can be changed without requiring significant changes in the other modules.

The proposal should describe the organization and recommended operating policies and procedure under the computer-integrated system. This description must provide documentation for the changes needed, and for the sources of the values that justify the computer technology investment. This is necessary to document the commitments made during the proposal development process, so that all can recall what changes needed to be made to capture the benefits.

The proposal should define in detail the implementation steps leading from the current environment to a fully computer-integrated firm. No organization should attempt to carry out the entire task at one time, for a number of reasons. The first is that the conversion is just too drastic a cultural change for an organization. To try to make it in one "bite," will guarantee failure. The second is that some elements of computer technology are not going to work well upon installation, and it's a lot easier to troubleshoot problems one at a time than lots of them at once. Third, breaking the total investment into a series of smaller increments makes the investment more palatable, and in addition, allows the installer to develop implementation skills and credibility as they go. Finally, computer technology is developing at an exponential rate; incremental implementation gives the firm a chance to take advantage of advances.

The proposal should define the actions to be taken in the first steps in detail, because it should act as the design guide for the first steps. It should define in detail the hardware needed, the software and operating system, the effort required to implement, and the changes required in the structure and organization of the firm. Specific suppliers can be identified, if appropriate. It should also include detailed operating budgets for all functions involved, reflecting the improvements expected.

This level of detail is appropriate for the first few steps only, however. The designers of computer integration must have freedom to alter as necessary and practical. In the later steps, less and less detail should be included, but it should also be made clear that the risk of forecasted improvements will become greater, also.

The proposal should conclude with the financial analysis of the proposed system including a long-range—perhaps 5 year—projection of the costs and savings leading to a summary statement of justification and payout period.

The guidelines described in "Justifying Computer Technology" should be reviewed carefully to help establish the case for this justification.

KEY ELEMENTS OF A CIM MASTER PLAN

John F. Snyder

General Electric Company

The CIM Master Plan: What It is, What It Isn't

Because the goal of Computer-Integrated Manufacturing is to computerize and digitally link the entire business environment, documenting such a plenary effort at the outset is vital not only to help sell the program to management, but also to define implementation principles and programs and to "drive a stake into the ground" as a basis for later measurement. Developing such a "master plan" defines the scope of the effort, without ignoring impacts and risks, so that all business functions hold common, realistic expectations of what is involved. As the individual sub-systems are presented for approval and funding over the years, the master plan provides the reference for management to see how that piece fits into the total picture and to gauge the impact on other sub-systems and the entire business if one sub-system is accelerated, delayed, or scrapped. The master plan, therefore, although dealing with technology, is not a technical document but a key element of a strategic and operational business plan. And it is a tool to educate how the business does operate versus how the business could operate.

Who Should Write the Master Plan?

Although a CIM plan requires the input and review of all business functions, a committee shouldn't wordsmith it. A committee can reach consensus on the broad concepts and structure of the document, individual technical and functional leaders can provide specific flesh to the skeleton, but one person with a multifunctional perspective needs to provide the balance and integrate the inputs into a consistent style and format. Failure to centralize the preparation of the document will produce a result that is too broad in some areas, too technical in others, lucid in some material, utterly incomprehensible to the layman in others. In other words, the plan document must reflect the same consistency, quality, and integration that CIM itself strives to achieve. The writer, therefore, needs to be more an organizer than a technician, more multifunctional in outlook than parochial.

Getting Started: Business Needs Drive CIM Efforts

Right upfront it's important to establish that a business doesn't undertake CIM because it's a hot buzzword or because some vendor has an offer you can't refuse. The only justification for CIM is to respond to the specific needs of the business's environment. Following the obligatory executive summary of the plan, the beginning of a CIM master plan documents the environment and needs. Is market share being lost due to inconsistent quality? Are customers demanding shorter cycle times? Why? Has the industry as a whole gone through some recent fundamental changes? Has the cost structure of the product changed? Has availability of material and people resources changed? This type of strategic questioning identifies the business drivers that the CIM plan will address and establishes a common perspective for all business functions. Business needs will also dictate the scope and duration of the CIM program.

Defining the "As Is" of the Business' Systems

The context of the business' needs begs the question of why current ways of doing things are not adequately responsive. The next stage of documenting the master plan lays out both the manual and automated systems being used by the business. This is a comprehensive look at all business functions, not just engineering or manufacturing systems. It's starting point is often the sales systems, carrying on through to quoting, booking orders, designing and engineering, resource planning, manufacturing responsibilities, through shipment straight to post-sales customer support mechanisms, with staff functions like relations, legal, and accounting hooking into the flow at the appropriate points.

A broad flow chart can be developed as support, detailed enough to give a flavor for the major mechanisms, activities, and interdependencies at work in each function, but not so detailed as to be confusing (unless, of course, that's precisely the point that needs to be made). At this point the presentation is factual, not judgemental. Some key decision influencer reading the plan may have been the architect of an inefficient "as is" situation. So it's purposes are to both inventory and understand the present for reference later and to educate the business functions as to how each fits into the current business operations.

There are at least two appropriate back-up documents for this section of the master plan which can be included as appendices:

1. A function-by-function list and description of the systems, processes, hardware, and software currently in use.
2. Current business data including organizational structure, facilities, workforce make-up, capacity, key financial statistics, etc.

This is not simply for current reference purposes. With changes and turnover occurring as quickly as they do these days, much of such data can be quite cumbersome to retrieve three to five years down the road in the process of auditing results.

So What's This CIM Stuff All About?

The next element of the master plan is pure education: defining the CIM concept. Too often a manager will think that having a flexible manufacturing cell (or having recently installed distributed numerical control, MRP II, or an expensive interactive graphics facility) is the same as having Computer-integrated manufacturing. A little missionary work on "islands of automation" vs. CIM is necessary here before proceeding to how CIM applies to the specific business.

Defining the "To Be" of the Master Plan

The stage is now set to lay out the details of the CIM program with a discussion of each sub-system. This is not simply a technical presentation. Indeed, highly technical descriptions, full of jargon and buzz words, are more likely to deter management from reading the document. Also, technical presentations will limit the interest of each business function to the sections pertaining only to their particular function. Since some technologies may be very dry, even arcane, the least that can be done is to make them reader-friendly.

Where to start? With CIM, the outputs of one sub-system provide the inputs for the next system, so start at the beginning of the business process: getting an order. From marketing systems, flow into the financial, engineering,and manufacturing systems that are triggered by the realities of the marketplace. There will be a number of systems that are generic across functions, e.g. office automation. Since they are ancillary but impact all functions, such sub-systems are among the three business activities of planning (e.g. MRP II, market forecast, business planning, etc.), execution (quoting, ordering, design, manufacturing, customer support, etc.), and measurement and reporting (financial and administrative systems).

The actual discussion of each system should contain a number of elements:

1. A description of the required functionality, including what systems and functions provide input and what systems and functions use the output.
2. Explication of the available technologies that can be brought to bear.
3. Anticipated new developments in the technology in the foreseeable future.
4. How the technologies can be applied within the specific business setting.
5. Typical benefits generated by the technology.
6. How the sub-system will be integrated with the business' other systems.

CIM developments are coming on stream quickly these days, so it's important to anticipate technologies that are currently beyond the state-of-the-art. They'll probably be available sooner than expected. Integration concepts need to be continually reinforced, too. One function may already have a state-of-the-art sub-system with which it is perfectly happy, but whose outputs are incompatible or useless to other sub-systems, missing opportunities for CIM synergies.

To complete the section, flowchart the anticipated "to be" model to summarize how the individual technologies fit together and for comparison with the "as is" model.

Establish Common Operating Principles

To smooth implementation, it is important for everyone to be operating under the same system assumptions and philosophies. Therefore it is important to spell them out. These operating principles and assumptions can never be completely documented, but major ones can be addressed. They are also wide ranging in their nature. For example, which sub-systems go in first should be driven by their criticality and potential benefits, so prioritize them. Another operating principle might be the policy to use off-the-shelf software exclusively with only a minimum of tailoring. Conversely, the principle may be to grow one's own systems from scratch. Another operating principle may be the assignment of overall responsibilities for various facets of the program. Should systems be rolled in one at a time or should a group of systems be implemented simultaneously? These and many other issues surface during the course of developing a plan and should be formally addressed within a logical framework.

One important operating principle is the avoidance of blindly automating the "as is." Since considerable resources of time and money are involved, it is vital to have all agree that time will be taken to analyze whether the current procedures and processes are really the best way to do things. Too often, a current mode of operation is an accretion of methods which individually made sense when first instituted, but which may make little sense in the current environment. For example, if a current manual system is providing flawed or sub-optimal outputs, automating it "as is" simply provides the same garbage more efficiently.

Timing, Resources, Benefits, and Risks

The final element of the CIM master plan is the one people are most curious about: what will be the effect and how will it impact me? To respond, what is needed at a minimum are Gantt charts showing each major sub-system and its associated major milestones; a forecast by year of financial investment required for each subsystem; an estimate of the man-months of human resources that are likely to be expended; anticipated return on investment and productivity increases; and a careful consideration of the risks involved of both implementing and, equally important, not implementing each CIM program element. Anticipating the effect and impacts might seem an overwhelming task, especially for a program that might take five or more years to implement. But remember, the initial masterplan document is intended to gain approval for the CIM concept and overall program, not to justify each of many many elements it is to contain. If the masterplan focus is minutely detailed in the beginning (rather than concept oriented), it runs the risk of being nickel-and-dimed to death by hair-splitters who can't see the forest for the trees.

Appendices Anyone?

Typical appendices are backup documentation of current systems and business situations, lexicons of technical terminology, reprints of CIM articles, etc. Appendices are also a good place to stick the detail that someone thought absolutely vital but which would have cluttered up the main body of the work.

How Much Effort Does It Take to Put the Plan Together?

This question can only be answered by the needs of the business and the intended scope of the program. One General Electric component, with sales of about one billion dollars, had two managers dedicated full-time for a year to put together a fifty-million-dollar, 10-year CIM masterplan. And that's not counting the scores of meetings of task forces among all the business functions that the initial planning process generated.

SELLING THE CIM PROPOSAL...

Donald B. Ewaldz and George J. Hess
The Ingersoll Milling Machine Company

Pre-selling is an Important Aspect

A key element of successfully proposing computer technology is pre-selling it as thoroughly as possible. Senior management must be fully convinced of the value and practicality of computer integration and eager to get to it before the work of developing the formal proposal is even begun. That can be difficult for several reasons.

An investment in computer technology is much different from an investment in a machine tool or other piece of equipment. One can demonstrate that additional product will be made by a faster machine tool. It doesn't take much imagination to discern payback from such an investment; it will come from the additional product. (The critical assumption is that the additional product has a market, of course.) Traditional justification procedures usually focus on measurable direct labor costs. The machine tool industry has had a strong hand in developing justification procedures to "justify" machine tool purchases. It doesn't take much imagination to come to grips with this.

Direct labor is a line item on the cost buildup. Unfortunately, those traditional procedures don't usually work well in analyzing other investments. (They sometimes don't work well in analyzing machine tool purchases, either.) But to an executive who is used to those procedures, it may be difficult to consider other cost factors. One must start early in the game to convince management that justification factors other than those traditionally used are valid.

Many executives and senior managers have little background in computer technology, and concomitantly, little basis for judging the worth of computer interaction. Certainly the easiest "solution" is to steer away from it. This implies some subtle training is in order. Not at the level of teaching machine language, or data structure, or system design, but training in how having ready and reliable information can solve management problems, can cut operating costs, and can result in a more responsive, more profitable firm. Fortunately, this effort is currently aided by a deluge of feature stories, articles, and non-technical books describing the advantages of computer integration.

There are usually few computer technologists about at this stage of the game, and unless the individual sponsoring computer integration has strong credibility in the science, it sometimes makes sense to use potential hardware and software suppliers to assist in the selling. Many suppliers have carefully developed persuasive sales presentations that can build a strong case for computer integration. However, it would be wise to review the presentations before putting them forward; material that is too technical obviously won't be of value, nor will material that is too obviously a sales pitch. It is also wise to limit the number of such presentations; the material becomes both redundant and murky very quickly.

The best approach to bringing the proposal to top management, under the circumstances, has two phases:

1. A traditional written format which includes all the normal elements of a request for capital. This should also provide ongoing documentation of the program and serve as a memory jogger for function managers during implementation.
2. A formal, carefully rehearsed live presentation; this will give the team a chance to field and answer any questions that arise, resolving them before they can become items which might defeat the proposal.

Some Criteria for Selling the Proposal

The previous text described the general format for the proposal. There are some additional caveats that must be observed in both the written and oral versions of the proposal.

The economics have to be "right." There are dramatic cost reductions and streamlining opportunities associated with computer integration. If they are not included, they probably have been overlooked, and the proposers are doing less than an adequate service to their firm. Costs that are unnecessary in a computer-integrated environment will continue, and the firm will be missing a major opportunity to become a tougher competitor. It doesn't make sense to spend money on "state-of-the-art" without gaining a good, solid, reliable return for it.

The package must be "politically saleable." A good share of the risk here will be overcome by the team approach to the proposal preparation. More will be answered by a good job of pre-selling the program. The rest must be addressed by sound understanding and recognition of the interest and idiosyncracies of the individuals who are going to pass judgement on the proposal. It will be less than wise to find undue opportunity in the accounting department, if the most senior manager reviewing the proposal was only recently the controller.

The benefits must be clear and credible. It's not unusual for technologists to become unduly optimistic concerning their technologies, and computer people aₑe also susceptible to this kind of thinking. It's not unusual for technologists to obscure their work in super-technical gibberish. Using the team approach to preparing the proposal will help avoid this, unless the functional managers become over-enthralled with technical jargon themselves.

The best presentation structure consists of three steps:
1. A brief statement of the costs and benefits expected, quantified in dollars and cents, including a translation of the income statement under computer integration.
2. A clear description of the proposed system, avoiding technical details (though being reasonably well prepared to address them if questions should be raised).
3. A description of the changes that must occur in the firm if the benefits are to be achieved, and something of the risks attendant in making the changes.

Who Presents the Proposal?

There are some obvious characteristics that dictate who the most appropriate presenter is, and none of them relate directly to function. All of them are specific to the individual firm; and most are just common sense.

The individual must have strong personal credibility with the individual or group of individuals to whom the presentation is being made. It's a lot more important for the presenter to be accepted as reliable than that he be recognized for skills in computer science, or any other kind of science, for that matter.

The individual must be clearly enthused and confident concerning the values of computer integration without being so enthused and confident that the proposal becomes irrational. Again, this doesn't mean he must be a computer scientist; it means he must be completely supportive of the data and ideas he is presenting, whether or not he fully understands them. It's likely the audience will be less versed, anyway. More computer technology proposals have failed because of too much "technology" content in their presentation than too little.

The entire proposal team should be present, showing enthusiasm and unanimity, and participating wherever appropriate. Just seeing so many functional managers in agreement about anything will go a long way in selling the proposal. It should be made clear that the proposal is a joint effort, not just the dreams of a single technologist, and that each manager sponsors and stands behind the proposed cost reductions.

JUSTIFYING CIM IS A SNAP

Kenneth W. Emery
General Electric Company

Long-term Commitment

The "selling" of CIM is not a one-step process nor its implementation. Commitment to CIM in one of operating philosophy, not commitment to a specific project. CIM means integration, and to integrate truly, every project must become a part of the overall CIM plan. The person who expects to fund one proposal and have CIM at the end of the project will be sadly disappointed. The decision to embark upon CIM really occurs in two steps. The first step is the decision to commit philosophically to the CIM approach, and the second is the funding of the separate CIM projects as they are formally proposed.

The first decision clearly lies with the CEO. Without the CEO's commitment there is little hope of breaking from the traditional practices in order to do things in new and different ways. The CIM process also entails the long-term commitment of personnel to the development of new approaches and systems which clearly cannot happen without CEO endorsement. Preparation of a CIM masterplan with clearly delineated objectives and milestones is vital to securing such commitment. The second type of approval, the financial funding for the individual CIM projects, is well-defined at General Electric and depends on project costs and benefits.

Financial Justification

The prime factor in project approval is financial justification. That is not to say that all of the reasons for implementing CIM can be reduced to dollars and cents. However, by examining the beneficial cost effects produced by each CIM project, financial effects can be determined; they usually demonstrate extraordinary savings. At General Electric, this calculation is performed by establishing the operational costs under the present operating conditions. The result is then compared with the operating costs under the proposed conditions. This projection is made as realistically as possible and includes such factors as actual shop loading, off-shift premiums, payroll benefit rates, inflation, etc. Depreciation is established for each category of expenditure and, along with other project expenditures, subtracted from the project savings to calculate the gross effects on the business' revenues. The effect of taxes is then calculated and combined with the investment cost of the project, investment credits, inventory reductions, and the tax benefits of depreciation to determine the total cash flow resulting from the project.

Cash flow can be evaluated in two ways. The first method is simply the traditional payback approach: a measure of the elapsed time from the initial project expenditures to the time when enough savings have been generated from the project to turn the cumulative cash flow positive. The second method is the discounted rate of return calculation. This equates roughly to return on investment and is calculated by determining the interest rate at which the present value of the sum of future net savings is equal to the present value of the net negative cash flow associated with project investment and implementation expenses. Using these methods, typical CIM project savings have ranged from a 35% to 45% discounted rate of return with a two-to-four year payback period (after tax, inflation adjusted basis).

CIM Savings

Some claim that CIM projects are difficult to justify financially. Some say that the savings associated with CIM are not of the same type they are used to dealing with. Everyone understands the financial impact of purchasing, say, a machine tool which reduces the direct labor necessary to make a part (i.e. savings = labor reduction x labor rate x number of pieces). Less obvious, perhaps, are the savings associated with reducing the cycle time of fulfilling an order from several weeks to several hours, particularly if this cycle time reduction does not include some reduction in the direct labor actually applied to fill the order. The fact that these savings are not of a traditional nature does not make them any less real or particularly harder to calculate. It simply means that we must examine the actual impact that these systems will have on costs and then quantify them. Typically, a CIM project will draw savings from many areas which must be combined to determine the full impact of the project. Too often, CIM savings are seen only as those associated with the automation of a process, in many cases not enough in themselves to justify the project. Over and above the automation aspect, most CIM projects have two qualities which are significant cost drivers: the speed at which the systems can respond and the benefit of accurate, consistent, on-line data.

Generally speaking, wherever the system brings improvement there is probably a manual effort currently being used to compensate for— or work around—the inadequacies of the existing systems. Find these manual, "coping," mechanisms and establish their cost to the business. Don't be timid about identifying any indirect activity to be eliminated and signing up the cognizant manager for the savings associated with their elimination. This can be accomplished in two ways. One is a proportionate reduction in personnel and the other is the value of the new work that the person can perform now that he or she no longer needs to perform the manual task. Examples include such activities as maintenance, indirect labor, expediting, key punching, numerical control programming, process planning, etc. Any activity that increases productivity will have a measureable cost impact on the business, just as inefficiencies caused by the slowness or inaccuracies of the present methods adds to present costs.

A second area that is often overlooked in calculating CIM savings is the impact that the system can have on inventories, particularly work-in-process inventories. CIM systems do things more quickly, so it follows that material spends less time waiting for something to happen to it. Less time waiting means less inventory. The traditional way of treating inventory savings is to determine the carrying cost of the inventory at some hypothetical time-value for money. But this anti-quated approach does not reflect the business' actual cash situation, and it under values the effect of the inventory reduction. The true effect of the inventory reduction is that of freeing up a lot of cash in a hurry, making cash available for other applications, namely implementing CIM projects.

There are some areas of CIM benefits that truly may be intangible. Will reducing the quoting or delivery cycles or improving quality result in increased sales? It probably won't hurt sales. Will freeing up your foremen so that they are managing people instead of filling out paper-work improve the efficiency of the operation? Hopefully it will. CIM will have a positive impact in these and many other areas. If you can get someone in responsibility to sign up for these positive results, include them in the financial justification. If not, leave them alone. Fortunately, most CIM projects can be readily justified based on the more concrete results previously outlined. It simply means taking the time to understand what actually is being accomplished and making sure that it is consistent with sound financial management.

FINANCIAL JUSTIFICATION OF FMS COMPARED TO CONVENTIONAL EQUIPMENT

Gary Lueck
Cone Drive

This paper will discuss Flexible Manufacturing Systems and Flexible Manufacturing Cells, contrast them to conventional machining approaches, and review problems which have arisen in financially justifying flexible systems and cells. It will also review some of the approaches used by Cone Drive in its plans for implementing flex cells. These cells have been an important part of our CIM program and, although the first cell is not yet operating, we expect to achieve many financial and operating benefits as a result of implementation.

FMS Defined

According to Edward J. Phillips, "Flexible Manufacturing Systems" (FMS) are engineered, computer-controlled, manufacturing processes that can adapt automatically to random changes in product design configuration, models, or styles. The system will always strive to optimize production output and work in process inventory." [1]

Flexible manufacturing systems are combinations of various machines or processes in order to complete a part or group of similar parts within the system. The system is often made up of several cells which manufacture various types of parts. The parts are generally received at the cell as raw material and leave the cell as finished parts.

Flexible manufacturing is unlike conventional machining because parts are not transported between departments in order to have the various operations completed. In other words, similar machines are usually grouped together in a conventional factory. Work-in-process is then moved between departments as the work is completed. In a flexible manufacturing system or cell, dissimilar machines are grouped together in order to minimize transportation and handling.

Advantages of FMS Over Conventional Machining

Flexible Manufacturing offers several advantages over conventional factory manufacturing systems:

1. Transportation costs within the factory are reduced. In many conventional factories, a part might be handled several times more often as the part is moved between various departments.
2. Work-in-process inventory is minimized. Because of fewer set ups and lower lot sizes, the total inventory is reduced.
3. Defective material is found sooner. This can result in:
    a. Lower scrap costs.
    b. Improved delivery performance.
    c. Less need for safety stock.
    d. Better overall quality and consistency of tolerance.
4. Improved lead times because of the lower scrap and because of the ability to economically run parts in the quantity required rather than the quantity needed in order to amortize the set-up costs.
5. Less space requirements because of lower inventory and higher machine utilization. This means lower initial costs and less ongoing expense to build and operate the facility.
6. Lower labor cost because of less setup time and material handling time as previously mentioned.
7. Less risk of obsolete material because unneeded parts cannot be justified based on set up costs. This may not be a factor because those companies with significant obsolete inventory problems will need to correct them before attempting to install FMS.
8. Operating more hours by running through lunch and breaks. It may often be possible to run machines without supervision for short lengths of time without the risk of a significant problem by having built-in safeguards.
9. Integrate with CAD/CAM to provide more accurate and more timely information to the Production Control and Manufacturing departments.

The above advantages of flexible machining can be contrasted to some of the problem areas under conventional machining which include:

1. Low machine utilization.
2. Excess material handling.
3. Expediting.
4. High inspection/scrap/rework.
5. Poor manpower utilization.
6. Long lead times for product changes.
7. High setup/limited flexibility.

Although these are serious problems which can't be solved overnight, flexible manufacturing can at least partially address these problems.

Flexible Manufacturing Cells

In contrast to the Flexible Manufacturing System which requires a large investment in financial resources and a management that has both the ability and the time to focus a great deal of attention, the use of Flexible Manufacturing Cells is a more practical alternative for many firms. Rather than developing a complete system at one time, Cone Drive has decided that the sequential implementation of manufacturing cells is more appropriate. We believe that this approach can provide many of the benefits associated with a flexible system with lower cost and less risk of failure. Under this philosophy, new cells will be installed as the previously installed ones become operational. By following this technique, we expect to be able to stay within our financial resources and be able to implement increasingly difficult cells as we gain in knowledge and confidence. This approach should help avoid the problem of overwhelming our financial and managerial resources. It should also help the firm develop more accurate forecasts of the time and money required to implement future cells.

Problems in Financial Justification of FMS

One of the main problems with justification of flexible manufacturing is that, almost by definition, the reason a company should investigate flexible manufacturing is because it does not have stability in its product design or in the product mix. Therefore, the products and the quantity of products to be manufactured are both unknown. Assumptions about either can be very difficult to make with much reliability.

"It is difficult to justify new technology by traditional cost benefit methods. The costs are current and easily measured. while the benefits are often realized in the future and not easily quantified." [2]

"ROI methodology assumes stability in the economy, technology, labor, and most important, the marketplace behavior of competitors—assumptions that have proven time and again to be false. In addition it stresses short run returns rather than long-run strategy...The difficulty lies in the disparity between the apparent ease of quantifying costs and the difficulty of quantifying benefits." [3]

The problem is not that Internal Rate of Return or Net Present Value techniques are inappropriate. Rather the difficulty lies in the fact that estimates of the financial advantages to be gained or lost to the competition are difficult to make. However, just because the calculations require estimates or assumptions that are difficult does not mean that the effort should not be made. Some of the more unusual areas to be addressed are:

1. Improvements in lead time. Because parts are run in smaller lots with less transportation and wait-time, customer orders can be completed in less time than if larger lots were manufactured in a more traditional setting. This can mean getting many orders that might otherwise go to the competition or would require large investments in inventory. The estimated marginal revenue for these additional sales or the inventory savings as a result of not carrying the extra inventory should be included in any investment analysis.

2. Improved delivery integrity. Orders are shipped when promised, not sooner and not later. Manufacturers with good delivery integrity can be counted on to supply products when needed. From the customer's point of view, there is little need for a vendor that can't supply the product when needed even if the price is right. The ability to supply a quality product on time is worth additional sales and profit that should be recognized in the analysis.

3. Future space requirements. Even if there is extra floor space at the current time, the added capacity does have a value that should be considered in the project evaluation. As Thomas Gunn notes, "Companies spend millions of dollars adding bricks and mortar to expand capacity and really capacity is just a function of parts per hour times the number of hours." [4]

4. Competitive reaction. It is likely that the competition will react

in someway. The most likely reaction might be to imitate. If done in haste this could lead to some poor decisions which may help you by leaving the competition in even worse shape. Although this requires some estimates and the use of probabilities, the expected advantage can be measured.

5. Future product changes. Improved designs and products can be more easily manufactured and marketed. This can mean increased sales in a market that has less competition. As with some of the other benefits, this might be difficult to measure but estimates and probabilities can be used.

6. Improved employee morale. This is usually overlooked but it is not unusual for employees to perform more efficiently after being given the tools to do the job better. It might be necessary to survey employees to arrive at a reasonable estimate of the increased productivity or changes in actual versus standard production rates after prior changes can be used as a guide.

There are some other difficult areas that require estimating after making assumptions about the market, the competition, or the workplace. These are generally more familiar and therefore more readily accepted by management. These include:

1. Inventory Reductions. This can generally be computed by comparing the planned lot sizes to the old runs. Remember that in a lot-for-lot environment there would not be any ending inventory for the manufactured part.

2. Scrap/Quality Expense. The estimated expense reduction might be 20% or 50% or whatever reasonable goal is established.

3. Obsolete Inventory Costs. A target percentage decrease such as 50% might be appropriate.

4. Labor Cost Reductions. This is the sum of the lower direct labor for setups, the reduction in material handlers and inventory control personnel, and labor costs while machines run unattended.

In summary, it is possible to translate the competitive advantages to be gained from Flexible Manufacturing Systems of Cells into dollar benefits. Although some of the estimates and approximations may be unfamiliar to many accountants or capital asset managers, the estimates are similar to all projections of the future because they involve uncertainty. It is better to include and attempt to quantify these benefits than it is to ignore them because of the problem of quantifying their value.

FOOTNOTES

(1) Phillips, Edward J., Flexible Manufacturing Systems: An Overview, Institute of Industrial Engineers, 1983 Fall Industrial Engineering Conference Proceedings, p. 639
(2) Computer Integration of Engineering Design and Production, Committee on the CAD/CAM Interface, Manufacturing studies Board, and Commission on Engineering and Technical Systems, National Academy Press, Washington D.C.: 1984., p.12
(3) ibid., p.35-38
(4) Thomas Gunn. "Computer-Integrated Manufacturing," APICS Operations Management Workshop, Michigan State University, East Lansing, MI. July 26-28, 1982., p.10

BIBLIOGRAPHY

American Production and Inventory Control Society, Computer-Integrated Manufacturing and Flexible Manufacturing Systems Seminar Proceedings, April 1985, New Orleans, LA: 1985

Bergstrom, Robin P., "FMS: The Drive Toward Cells", Manufacturing Engineering, August 1985: pp.34-38

Computer Integration of Engineering Design and Production, Committee on the CAD/CAM Interface, Manufacturing Studies Board, and Commission on Engineering and Technical Systems, National Academy Press, Washington D.C.: 1984

Davis, Jim, "CIM—A Competitive Weapon," Manufacturing Engineering, August 1985., pp.16-19

Gunn, Thomas, "Computer-Integrated Manufacturing," APICS Operations Management Workshop, Michigan State University, East Lansing, MI, July 26-28, 1982

Phillips, Edward J., "Flexible Manufacturing Systems—An Overview," 1983 Fall Engineering Industrial Conference Proceedings., pp. 639-645

Stage 1
Introduction to CIM

Stage 2
Preparation for CIM

Stage 3
Program Plan for CIM

Stage 4
Implementation of CIM

MANAGING IMPLEMENTATION PHASES OF CIM

Paul W. Brauninger
Cone Drive

If we agree that CIM is a concept that involves the organization as a whole, then it seems logical to assume that the person responsible for the total organization should also be responsible for the CIM program. In essence this is correct, but the nuts and bolts of implementation are another story that necessitates the delegation of authority from the Chief Executive to another individual who has more time to devote to the intricate details of implementation. The Chief Executive must select an individual who understands the needs of the organization, is informed, and who can make a decision at a critical moment.

The CIM Project Manager

Since CIM involves the total business, the CIM project manager should be familiar with the long term business plan, have a basic understanding of the different functions in the business, be a person with an attitude that believes that "change is good," and most important, be a leader who is able to get work done through people.

This person must be someone who can focus on the objective and see it through until the end. However, in the case of CIM, there is no defined ending point. Instead there is often an unending stream of projects. Computer-integrated manufacturing is an ongoing objective that links all areas of business to develop a fully automated manufacturing operation that: reduces inventory investment, permits economic production of parts in lot sizes of one, and reduces product lead times to customers while becoming more cost competitive and maintaining or improving quality.

The skills required for successful integration are not only the normal technical skills in engineering, accounting, or manufacturing, but also effective management skills. In addition, the CIM manager must have the faith and confidence of the chief executive because he will make decisions affecting the long term viability of the business. The

CIM project requires an individual who is a decision maker and will take responsibility for the project. In most cases, a wrong decision is better than no decision at all. Hindsight is often very useful after the project is done. However, second guessing during the project often serves only to divert management from their desired objective.

Communication

It has been our experience that equally important to having the right person responsible for the project is the establishment of a good communication system. When implementing a CIM program, change is a required ingredient. In order to have change accepted, it is usually mandatory that all people or departments affected by the change "buy into" the new idea, establishing a sense of ownership in the project and the eventual outcome. An effective communication system provides an atmosphere that when a change in one functional area affects the work environment of other areas, the benefits are communicated first to the other affected areas.

It is also crucial that the basic concepts of the total CIM plan be communicated to all employees involved. Regular progress meetings should be held to review progress made-to-date, to identify potential problems, and to assign responsibility to individuals for accomplishing new tasks. An important task of the project leader at these meetings is to be certain that everyone knows what new decisions have been made and how each department will be affected by these decisions.

To ensure the success of a CIM project, a written milestone plan is required. When implementing this plan one must develop and maintain an attitude that says delays are considered unacceptable. Don't accept excuses and reasons for not meeting the schedule. Find out what is required to reach your objectives and make sure it gets done. Developing this attitude within everyone involved is critical if you want to reach your goals. Even if only one or two individuals begin to fall behind in their areas of responsibility, the CIM program could be in peril.

Celebrations and Old Habits

Two other areas must be addressed when managing the implementation of a CIM program. First, the CIM project manager must ensure that the syndrome of "hooray, we've accomplished this task so let's stop a minute and rest" does not develop. Celebrations are great and often necessary, but don't fall into the trap of stopping, because it just gets harder to get started again. Don't say "well, we've accomplished this much, now let's wait awhile before we do the rest." Resting on past achievements will not help achieve new goals. Secondly, the CIM project manager must ensure that a constant vigilance is maintained to keep the program intact to prevent slipping back into old habits.

Old habits die hard. When a crisis occurs, people have a tendency to go back to their old methods of doing things. When something goes wrong or isn't turning out the way it was planned, it is necessary to step back and try a new approach rather than allow people to revert to their old ways of doing things.

In summary, the key to managing a CIM project is the management of people. Rarely can the failure of a CIM project be traced to hardware or technology problems. The problem usually lies with inadequate management of people.

The CIM project is typically a significant financial investment for a business and, in many cases, the successful implementation could mean the difference between success or failure of the business. Because of this, it is imperative that the CIM Project Manager be hand-picked by the Chief Executive for exhibited superior working knowledge and management skills. Not only must the right individual be selected for implementation of the CIM project, but top management must also make it evident that it is visibly behind the program.

WHO SHOULD RUN THE CIM SHOW?

John F. Snyder
General Electric Company

Computer-integrated manufacturing? Computer systems organizations commonly are a component of the financial function. Let them take on CIM? Perhaps no. It's true that historically Finance often evolved into the landlords of the mainframes and the home of the analysts and programmers, but the cross-functional interdisciplinary demands of CIM implementation beg a new structure.

It is also common these days to see specialized systems organizations dispersed among a wide variety of functions. Finance may hold sway over most of the mainframes, but as mini-computers and higher-level languages were developed, everyone got into the systems act. Engineering formed components to implement CAD (Computer-aided design) and CAE (Computer-aided engineering). Manufacturing built up their systems troops for such disciplines as automated process planning, distributed numerical control, MRP II (Manufacturing Resource Planning), etc. So too with the specialized needs of the marketing function and relations. Such fragmentation of system resources, often resulting in highly independent and parochial fiefdoms, can impose formidable blocks to implementing CIM successfully.

One solution a number of businesses are evolving to is a centralized CIM structure organized along the lines of four basic activities in the systems life process: conceive, design, implement, and support (see *Figure 3*). The CIM component, like the other functions, reports to top management and contains not just systems expertise but expertise in all business functions. By having such centralized talent and by being a component independent of the other functions, the CIM organization can perform the integrating role that is at the heart of the CIM concept. Here's how it works.

Process:	Conceive ──▶	Design ──────▶	Implement ──────▶	Support
Component:	Advanced Systems Design	Tech Systems & M.I.S.	Sub-system Project Teams	Operations & Systems Support
Tasks:	Planning & Integration	Systems Development	Installation	End-user Support
Expertise:	Multi-functional	Functional	Project	Systems

Figure 3
CIM Process as Structure

Conceive

Since implementing CIM involves developing and integrating a host of sub-systems, one component is charged with maintaining the "Big Picture." This component, call it Advanced Systems design or CIM Planning and Integration, if you will, not only conceives of and seeks out new computer technologies to fold into CIM, but also develops and monitors the major element of the formal, long-range strategic CIM master plan, particularly the anticipated effects and benefits to the business as a whole of all the CIM efforts. The component ensures that the sub-systems being developed fit into the grand scheme and integrate effectively with each other. Although technical expertise is important in the component, of equal importance is a high level of multi-functional expertise and sensitivity, since this group's charter often necessitates facilitating the resolution of the conflicting needs and priorities of disparate functions.

Design

Once the broad objectives and strategies of the CIM effort are defined, two components can see to the development and technical integration of the required sub-systems. One component is called Technical systems and the other Management Information systems. Within Technical systems are the experts who used to work for Engineering developing Computer-Aided Design, for example, or for Manufacturing on process planning. Here too are those developing office automation applications, professional productivity tools, and communications and numerical control programming technologies, among others. The Management Information System component is where the analysts and programmers who used to work under Finance wind up. Their charter remains the development of business rather than technically oriented systems. The common need in both these groups is for a high level of functional expertise as well as technical depth in such fields as group technology, databases, product design, communications, hardware, etc.

Implement

Once the CIM sub-system has been developed and debugged, including the linkages that communicate with the other sub-systems, someone has to physically install the system and train the new users. This task falls to components set-up for the life of the specific project. Here you'll find project managers working with Plant Engineering to pull cable, install computer rooms, etc. Here too you may find the component engaged in setting up an MRP II system for the business.

The implementation teams need a project specific expertise coupled with a site specific expertise to be effective. Although the end-users of a sub-system have been deeply involved in the Conceive and Design phases of the project, the actual Implementation phase needs the most delicate and skillfull handling by the implementation teams to minimize disruption due to changeover to a new way of doing things.

Support

Once the CIM sub-system has been installed, it remains axiomatic that the best of systems is worthless if they are not used. Through the life of any system, user support is absolutely crucial, crucial enough to warrant the special attention of a dedicated component. Few things can infuriate a user more than to phone "Joe Hardware" with a problem and have Joe say "Sounds like a programming problem to me. Call Jane Software." Of course, when the user calls Jane Software it sounds like a hardware problem to her. The component could be called "Computer Systems Operations and Technical Support." It would provide a single focus for supporting end-users. If a user has a problem with his graphics system, for example, he simply calls Frank Graphics, in charge of supporting that system. Frank figures out what's wrong and the appropriate resources for remedying it. This component, on-call 24 hours a day if need be, requires expertise in the specific system. A member of the original sub-system implementation team is often the best candidate to move over to the support component.

Reporting and Control Mechanisms

Whatever structure is used to implement CIM, control of the program is absolutely vital; every single business function is involved in some facet of the project. To keep track of the myriad of activities, several levels of planning and control are necessary.

At the highest level is the "master plan." Here, philosophy, operating principles, programs, broad timetables, and resource requirements are spelled out to provide a specific yet flexible road map to get from "what is" to "what will be." Whether the master plan is for a 3, 5, or 10 year effort, the broad overview is important. This master plan needs to be updated regularly through the life of the project to reflect changing technologies and business needs.

At the more operational level, however, detailed schedules and resource allocations need to be determined. On an annual basis, management from all functions must sit down and work out monthly milestones and assignment of specific responsibilities for all facets of the project, including commitments for manpower support from each function for each project. These detailed plans provide the basis for top management's monthly review of progress. Although this process of achieving consensus among the different business functions on resources can be quite difficult, but it must be done to achieve the synergy that is the power of Computer-integrated manufacturing.

MEASURING AND EVALUATING CIM RESULTS

Donald B. Ewaldz and George J. Hess

The Ingersoll Milling Machine Company

Establishing Baseline Data

Getting the benefits of computer integration is going to mean that the organization of the firm, and procedures, and policies that guide it are going to change. Measuring those benefits means establishing an as-is configuration of the firm. The as-is configuration should define the current operating environment in terms of elements of the income statement—cost-of-goods sold, the responsiveness of the firm to customer orders or order changes, cost of product quality, and a table of organization describing the real organizational structure, instead of the one the annual report shows.

The as-is baseline should specifically identify the elements of the overhead accounts and their actual expenditures at some known level of business. These accounts are without doubt going to be the most important source of cost reduction. Unfortunately, values are traditionally measured against unit product cost in terms of direct labor cost, material cost, and overhead cost. Cost analysis has developed along the lines of evaluating product cost at that level. Little effort has been directed to measuring, or even sizing overhead costs. The value of overhead cost is usually based on an absorption constant which may be updated infrequently — sometimes only annually. If the absorbtion constant times the quantity of the independent variable against which it is being absorbed doesn't match the actual expenditure in the overhead pool, either too large or too small, the mismatch is accommodated by applying a "variance adjustment" at the end of the accounting period. The variance adjustment matches the amount of overhead "absorbed" to that actually spent. The variance may be negative; too much overhead has been absorbed. In such a case, product cost has been overstated; an adjustment is made to reduce the apparent cost of product. The need for a variance adjustment might be the result of "over absorption"—more labor hours or dollars being applied, or by spending less than budgeted—or the opposite of these if the variance is negative. Or, it may have occurred because the firm actually achieved the gains planned from applying computer technology, but did not recognize it financially until the end of the year, and then only in an income statement account in which it was lumped with other variances.

Consequently, the gains made because of streamlining the firm through computer integration can be actually achieved, but not be visible because of the idiosyncracies of the accounting system. It's extremely important that the means of tracking and validating the predicted improvements be established as a part of the proposal. Otherwise, the successful implementors might find themselves the victims of overruns in other budgeted costs, variations in demand, or poor performance in other areas. Clearly, the accounting function must be included as an important part of the overall proposal planning and tracking of computer integration. Otherwise, it may be extremely difficult to detect if any gains, even dramatic ones, have been made.

It is much easier to deal with the overall overhead pool than the application of it. It's too easy to lose real gains in reducing overhead, be it lost in reduced labor hours or dollars—both good, but having the effect of increasing the overhead rate. The improvement forecast should be the difference between the as-is base line overhead pool, adjusted for differences in activity levels, and the overhead pool in the environment of computer integration.

Since reduction of inventory at all levels will also be an important benefit from computer integration, every effort should be made to identify how much inventory is viable, and how much is actually dead stock. Computer integration is not going to make unsaleable goods saleable; materials produced as overage on Korean war military contracts will not suddenly become valuable because the firm's operating structure has changed. (Computerizing inventory records, of course, will make it easier to identify what's in inventory, how long it has been there, and where it is. It will help the managers of the firm make sound business decisions; it won't guarantee all decisions will be sound.) The as-is baseline must identify what inventory has a market, and the turnover rate of each item, so that the target for inventory reduction will be a realistic one. If there are 12 years' supply of a particular product in stock, it is unlikely it can be reduced a great deal in one year.

Data Collection and Progress Measurement

Traditional audit systems usually deal with elemental variable product costs, evaluating the impact of an investment on direct labor and materials cost. Since computer integration focuses on the cost elements traditionally considered fixed, the impact of computer technology on the firm's operating costs must be measured by changes in those accounts. Measuring the cost elements of individual products is not likely to show the true impact of computerization:

1. Direct labor costs may not be affected at all, at least not at the level of cost per part.
2. The overhead absorption rate will probably be wrong; it may not have been changed at all from the non-computerized environment, if the accounting and cost accounting functions haven't been included in the proposal team.
3. Impact on materials cost is more likely to be in the area of changes in purchased quantities, and probably won't affect "standard materials cost" at all, if, as above, the accounting functions haven't been included in the proposal team.

Variance Analysis

Variances should be measured by comparing budgeted expenditures from the computer integration proposal against actual expenditures. The comparison should be at the account line item level, not at the level of the cost of an actual unit of product. This eliminates a possibility of distortion from performance variations in other areas, against demand changes, and all the other variables which can impact individual unit costs. It also has the advantage of pinpointing variations. If a line item of a particular account overruns, the reasons for such overrun can be explored, and corrective action taken. It's not unusual to find that managers have found some new tasks for individuals planned to be idled with the advent of computerization. The best solution for this is to treat retention of an idled individual like a new hire; force the manager to go through the same justification process to keep a person as to add one. This will make actually achieving planned cost reductions considerably more probable, or at least make public the reasons for not meeting them.

Improvements in responsiveness are somewhat easier to track. It's not difficult to audit the time required for each step between order entry and final product shipment. If there is a great deal of activity, however, there may be wide swings in performance. The best answer is to deal with averages over time, which should be one of the products of the as-is baseline development.

Inventories constitute somewhat of a different problem. If the firm has done a good job of breaking inventories into classifications according to viability and turnover rate in the as-is baseline formulation, it won't be difficult to measure progress in reducing inventory levels. If inventories are being treated as broad levels of "work-in-progress," raw materials, and finished goods, it may be next to impossible to see any discernable slimming. As discussed earlier, reduction targets should be realistic, and directed toward specific inventory items or accounts, and managers should be very firm in assuring the targets are met.

Corrective Actions and Adjustments

Computer technology and the integration of all the functions of the firm offer perhaps the most important opportunity for managers to streamline their firms since the industrial revolution. It provides firms the chance to compete with offshore competition whose advantages include lower labor rates, favorable treatment by their governments, and favorable exchange rates. Unfortunately, long-standing traditions are tough to overcome; operating policies long in place (and whose original reason for being may have long since passed) are difficult to overcome. The changes that were so obvious and practical during the proposal phase often become a lot more difficult to carry out in implementation, not because of technological or task constraints, but because they mean displacing people, and that's always difficult. They mean changing policies, and people dislike policy changes, even though the original policies themselves may have been distasteful and constraining. However, capturing the opportunities that computer technology offers means change, however uncomfortable it might be in the short term, and the senior managers and executives of the firm must insist on meeting the budgets developed in the proposal phase. The alternative is to make the investment and have it fail financially, however technologically "swishy" it might be.

The planning done in the proposal preparation phase should have provided a step-by-step, sequential plan for computer integration. The senior management of the firm are responsible for assuring the budgets established at that time are met, and that the introduction of computer technology actually achieves the dramatic kinds of cost reduction and organizational streamlining of which it is capable.

SUMMARIZING AND SHARING THE CIM EXPERIENCE

John W. Pearson
AT&T Technologies

An obvious benefit to be gained from the experience of implementing CIM technology into a business is that of further spreading cost-saving ideas and productivity improvements into other segments of the business and into other sectors of the economy. In some instances, a business will consider certain benefits derived from CIM implementation to be proprietary, but there are some broad concepts of such developments that could be shared for the overall benefit of society.

Some thoughts are expressed herein on how experiences of implementing leading-edge manufacturing concepts can be shared. It is always appropriate for any business to carefully review global aspects of its business operations and weigh the potential for sharing and exchanging ideas with other similar businesses. The ultimate goal of such exchange is to develop additional possibilities for further improving operations or productivity.

Techniques for Reporting CIM Successes to Personnel

Getting Employees Involved. Whether the effort to integrate CIM into a business is for a new facility or to upgrade operation of an existing one, an opportunity exists to involve all personnel in the activity. This will produce more than just a cursory benefit. If personnel are involved and are part of the process of improving the business, they will be more receptive to changes that result from the implementation and will contribute to the definition process as well.

The value of employee participation has been proven in recent efforts to involve employees in such groups as Quality of Work Life teams and similar groups. Employees perceive themselves as part of the management of the business, and the business profits as employees offer ways to reduce costs and refine business operations.

If employees are involved (encouraged to suggest or offer ways to automate or improve operations) in planning activities and informed of progress made during development, they will respond favorably to the changes since they perceive themselves as part of the business rather than working for the business. An effort should be made to periodically brief employees of progress in implementation that affects their role in performing business activities. It would likewise promote morale and stimulate employee participation to hold periodic reviews of the overall planning and progress made toward CIM implementation.

Employee participation may be viewed as a key point in producing new ideas in improving the business and smoothing the path to changing operations with minimal resistance. Just as allowing operation organization personnel to participate in the planning processes, it is crucial that persons expected to use CIM technology in daily operations feel they are part of the activity. One of the best ways to attain this goal is to allow such persons to be informed of plans and to comment or suggest areas that may be further improved or incorporated into the plans.

Of course, the team must maintain control of the planning process since many employees may not have first hand information on the global needs of the business. Care should be taken to encourage employee interest while avoiding situations where the planning process is impeded by excessive reviews or unproductive discussions.

Internal News Reporting. The process of keeping employees, as well as staff, informed could be augmented by several conventional methods. Typical activities such as weekly or monthly status review meetings could be scheduled. Personnel to be invited to such meetings would include operating section chiefs, department heads, engineering, production control, and other persons that directly participate in the process of planning and operating the business.

Another common method for distribution of news would be to publish an internal document such as a newsletter or bulletin. Such documents may include the usual topics of cost reduction achievements, benefits, employee activities and accomplishments, changes in business goals, and other items of interest to employees. The newsletter can just as well serve as a medium for review of new ideas to be implemented in the business (CIM technology, automation, etc.) as well as progress achieved toward those goals.

Periodical Business Results Meetings. Many businesses have frequent meetings among key personnel to report and discuss current results from business operations and to present near term-expectations of the business. This also may be a suitable time to review and report information regarding CIM activities.

In meetings such as these, it is generally appropriate to make brief, summary remarks about any topic since the intent is to provide a review and not to elaborate. Only a quick review of new developments or plans should be presented here. More detailed presentations or discussions should be deferred until a special meeting can be arranged since most persons invited to a "results' review are receiving information affecting their role in operating the business and should not be kept away from the job unnecessarily.

Encouraging and Allowing Technical Publications. In a typical business, there are individuals that may be inclined to write articles suitable for publication. This practice can be used for benefit of both the business and the individual if it is properly guided. Management should consider benefits to be gained through encouraging such participation. Some attention and notoriety can be good for the business and may lead to further development of cost saving ideas that can be implemented into operations. Responses from readers may also lead to exchanges of information that can improve business or avoid mistakes that could be built into a process in error.

In stimulating technical publication activity, care must be taken in making sure the subject material is adequately covered to make the document interesting. Good technique must be encouraged while also allowing the author to be expressive and informative.

In the process of developing documents for publication the writer must be careful to discuss enough information to inform and stimulate the reader, but not reveal the details that may be of a sensitive nature to the business. This is a common practice that may be observed in many technical discussions in trade journals. The author should review the subject from a generalized view but not describe details that reveal actual implementation of an idea or process. It is quite possible to review a concept or application and not even touch on the actual implementation details. This method can allow the business, and the author, to realize the benefits of publication and retain cost benefits of the application in a competitive business situation.

The process of publication can serve a number of purposes for the business. Innovative concepts can be introduced to businesses in general and help to stimulate increased productivity in the economy. The response and feedback from readers can lead to exchange of additional concepts and ideas that can in turn further fuel the reduction of costs and improvement of business operations. The individuals who develop the publications, in turn, will likely be stimulated to be more productive and will be a greater asset to the business.

Employee Training Programs. Training programs provided within a business structure offer further opportunity to stimulate employee involvement and interest. A well organized training session will provide needed exposure to individual operations that the employee will participate in, but can also expose the employee to current and planned enhancements in the business that will affect or involve the employee.

The training process can be used to create and stimulate a climate of receptiveness to a changing business. There is opportunity for the business that is receptive to using training as an additional tool to motivate employees.

Rewards for Technical Accomplishments. Since CIM and other high-tech concepts are costly and require considerable effort in planning and implementing, it would be fitting to recognize the accomplishments of the persons completing the efforts.

Mention of team accomplishments in internal newsletters or other publications can accomplish the dual role of informing employees of new or continuing activities and will further encourage productivity of the persons performing the effort. Care should be given to recognizing all personnel for their contributions in reaching the CIM objective. The actual success of the implementation is a result of the combined efforts of management, engineering, operating and all support persons, and they all should share in the success of any major effort such as CIM development.

Techniques for Reporting Successes to Others

The actual tools of reporting any major successes appear in many forms. A few of these are listed here as thought starters:

1. Encourage contributions in technical publications.
2. Encourage participation in technical seminars and presentations.
3. Encourage participation in technical societies and in presentations in programs.
4. Solicit exchange conferences or information interchanges.
5. Allow public or exchange tours.

Participation in professional, industrial, or trade seminars can be an exciting experience for both the business and the individuals involved. Preparation and presentation of a paper for a seminar can encourage the author to view the subject from the listener's perspective and may provide insight for further enhancements in the business processes.

The presentation made at a seminar will almost always lead to sharing of ideas from other technical and management persons in similar business and persons met can be a valuable resource for ideas.

MOVING TO THE NEXT CIM OPPORTUNITY

Paul W. Brauninger
Cone Drive

It would seem that after being successful in implementing Computer-integrated manufacturing (CIM) in one area of your organization that the next implementation phase would flow easily. If you are fortunate enough to have developed a CIM plan that encompasses 3, 5, 10 years in the future then you probably know where the next step is. But if you haven't developed a plan or the markets that your firm participates in are constantly changing, you may have to rely on other methods to point you in the right direction.

The Business Plan

The first area to look for direction is in your business plan. The business plan should help you identify the key areas that have to be changed or that need additional work done in if you are to meet your plans for the future. If you have not already developed a CIM plan to go with your business plan you probably should make that a high priority item. And if you already have a CIM plan it should be reviewed along with the business plan to ensure that the CIM plan meets the needs of the business plan.

A major factor in providing direction in where to go next results from analyzing the needs of your organization, and determining how receptive the area you have selected will react to change. If there are two areas in your organization that are equally critical to your business success and one can't wait to install a CIM program and the other doesn't want to even hear the word change, then the decision is easy. The hard part is when the group wanting nothing to do with CIM is an area that must be changed to achieve your business plans.

Integration

In looking into the next CIM opportunity, it is important to consider the issue of integration. The ability to integrate one system with another is the cornerstone to having a total CIM program at a facility. While there are exceptions to every rule, deviations from the integration concept invite serious future problems.

Focus should also be directed toward establishing basic systems that provide accurate data bases that will be needed by other systems. These basic systems should be in place prior to undertaking projects to automate other areas. This concept is especially important in implementing the second CIM opportunity, because there may be pressure to move to an area that cannot be done until some other problem or area has been addressed first. Remember that a failure or setback in one task can seriously jeopardize the viability of future projects or the whole CIM plan.

Momentum

To help ensure the success of the next CIM opportunity, the next stage and timeframe of the CIM plan must be defined before the previous CIM project has been completed. Momentum is easily lost and hard to get started again. Since the greatest factor affecting success of these projects is people, you need to keep the excitement and momentum going. The organization will have developed a sense of confidence in completing the first stage of the program. You need to utilize the enthusiasm and confidence on the next opportunity.

In the process of completing the first phase of your program you will probably have developed some very talented and motivated individuals. If there is a delay in starting the next phase, boredom may set in and these individuals could lose their enthusiasm or they could be reassigned to other areas of the organization.

Be prepared to work off of prior successes. If you have generated a great deal of enthusiasm in one area, see if something else can be done in that same area. But also be vigilant for a new attitude in areas where you may have been initially rejected.

In some cases, projects which are insignificant by themselves can contribute a great deal to the total CIM program. At Cone Drive, we found we accomplished a number of tasks others had not accomplished simply because we did not know it couldn't be done. Be watchful to ensure that the basic guidelines are being followed, but allow people to use their creativity to solve problems and provide input into what task should be done next.

Another item to consider. Try to implement CIM programs in areas where technology is stationary for a moment. But be cautious of being caught in the syndrome of not doing anything until the ultimate solution is available. Keep in mind that success in installing a CIM program is the ability to improve your company's position in its marketplace, and immediate action may be required. Unless a decision is irreversible because of the size or the direction of the commitment, it is often better to make any decision than it is to stay where you are. Be careful however that some decision which seems relatively insignificant doesn't determine the answer to some larger question. For example, the choice of an operating system or software may require certain hardware that is not being looked at or considered.

GETTING THE WORD OUT ABOUT CIM

John F. Snyder
General Electric Company

Communicating CIM progress and accomplishments falls roughly into two categories: internal and external. Internal communications involve not simply the formal reporting mechanisms for management to monitor progress, but also letting the business as a whole know what's going on and how each function fits into the picture. External communications, on the other hand, get you into the realm of interpreting your efforts to such publics as professional and educational groups, industry leaders, and the full range of media. Here are a few tips based on the experiences of a business that successfully integrated its systems from the point of quoting to a customer through to shipping the finished product.

External Communications

After you've spent prodigious time and effort to develop and implement CIM technology, it's very gratifying to be able to toot your own horn a bit. If you've achieved a quantum leap in productivity through the project, business realities may force you to keep mum so that the competition doesn't catch on, especially if it's a technology that's relatively easy for them to duplicate. Even in that situation, however, the benefit of publicizing your accomplishments and improving viability in the marketplace may outweigh the dangers of letting competition in on what you've been up to. But let's assume that management has given the go-ahead.

The first tip is to make sure you have something to show. Sure, the most spectacular CIM projects are often software oriented—a bunch of bits and bytes flying around inside a box. Even technically sophisticated audiences like something they can look at. And you can't really bring the media in to look at all the people who aren't there any longer because of the technology. Besides, doing that is probably in poor taste anyway, particularly if you are dealing with the local press. Getting good coverage really is helped by having something demonstrable that can be understood even by a layman. If you issue a press release to the local Daily Bugle telling about your big investment in MRP II, be prepared to get a phone call from the newspaper asking to come over and take a picture of your new MRP II robot.

If you are serious about getting good coverage for your accomplishments, also be prepared to spend some money to do it right. You'll be looking at getting articles into industry and technical journals as well as newspapers. There will be speeches to give at professional conferences, educational institutions, and Chambers of Commerce. If your CIM program is a showcase site, you'll have significant demand for tours of your facilities by groups of a wide range of technical sophistication. Providing quality support for such activities is vital to the image and message that is to be conveyed. Expect to expend resources on videotapes to support presentations, high quality slides (not overhead transparencies), first class industrial photography, and ghostwriters for speeches and articles.

Ghostwriters? A good technical writer is worth every penny. At a recent conference sponsored by the National Science Foundation, in Stuttgart, educational, business, and government leaders met to discuss the curriculum needs for a "CIM engineer." There was talk of the need for a "renaissance engineer:" one adept in systems, computers, manufacturing technologies, engineering, and finance and marketing and project management. But a heated issue turned out to be for educational institutions to start turning out engineers who could write intelligible English and express themselves well on their feet.

Media coverage of the CIM program should not be restricted to the technical aspects. Since CIM's ideal is to integrate all the business' functions, non-technical media can be served as well. Marketing associations and publications will be interested in how the technology will impact their competitiveness and aid sales efforts. Employee relations interests will be open to news on the impact of the technology on the work force. Financial groups want real-life examples of how to justify, measure, and track high technology projects. Management associations look for éducation on how one plans for, implements, and controls CIM activity.

Quantifiable financial benefits to the business of broad media coverage of the CIM program can be elusive. There will be consensus that it is good public relations if the effort is well-planned, targeted, and consonant with the broader goals of the business.

Internal Communications

Say you've just made a big digital leap. Say you've finally developed a link that takes design data from engineering, instantaneously develops a tape image for distributed numerical control, and sets the robot to assembling or the machine tool to cutting. What's the first thing to do? Throw a party, of course. But who should you invite? This apparently frivolous question is not trivial.

Effective internal communication about CIM is essential to maintaining momentum and easing future implementations and enhancements. The formal communications mechanisms that management uses to monitor and control are vital but so are the informal communications which educate all functions to the pervasive effects of CIM. The answer, then, to the question of who you invite to the party is—every function. Someone in finance surely had a role to play in originally evaluating the economic feasibility of the project. Employee relations surely should have been involved at some point, particularly if there were labor take-outs. Marketing should be aware of what the project accomplishes in reducing cycle time to react to customer's short cycle needs. And wasn't there at least one secretary who let you know top management's mood the day you had to go in and report a major snag halfway through the project?

Education and climate-setting activities involve internal publicity and recognition programs for extraordinary effort. Recognition involved the people designing and implementing the system, of course, but recognition of the end-user as well as those peripherally in the project is just as important to foster teamwork and gain system acceptance.

Education and climate-setting can be subtle too. Consider this real grassroots implementation tactic. It was decided to install factory-hardened personal computers at each machine tool work station in the shop. These intelligent terminals would be used for distributed numerical control (DNC), material and job-status tracking, and communications for the operator to summon support functions to the work station like the foreman, machine repair, movemen, etc. There was real concern whether the workers would accept this on-line, real-time system or whether they would see it as a "Big Brother Is Watching" ploy by management. A good deal of time had been spent educating the workforce to the business imperatives of automation but would they be intimidated by ignorance of how to use computers? Formal training programs were established, of course, but a major factor in widespread acceptance turned out to be which operators got the computers first. Informally it was identified which workers had simple computers at home...or video game attachments for their TV sets...or were addicted to PACMAN. These workers were hooked-up first. They picked up the user-friendly system very quickly and were soon showing its capabilities off to their cohorts. The response of the rest of the workforce quickly became, "How soon can I get my computer?"

The lessons of this example and the question of who to invite to the party are two-fold. First, involve as many as possible in CIM education and recognition. Second, do not let your imagination be bound to traditional modes of communication. Newsletters are fine, but word-of-mouth and peer examples are more powerful.

On the other hand, word-of-mouth causes severe problems if it's the principal way of communicating project status and results to management. The other extreme, of minutely detailed project schedules replete with PERT and GANTT charts offer more visual comfort and are powerful tools, maintainable on personal computers. But the limitations of current commercially available scheduling packages have to be recognized. Too often they are not "robust," not flexible enough to the rapid, dynamic changes associated with many high-technology projects, and the often tacit assumptions of their algorithms can lead one down a primrose path. If the approach is too detailed, it's entirely possible for a project to be completed before all its nodes and interdependencies have been defined. Monitoring mechanisms that emphasize practicality over elegance are preferable.

The middle ground of monitoring progress identifies what management needs to know and how often it needs to know it before institutionalizing the formal mechanism. The CIM overall plan will provide the guidance in gaging these questions. A well documented, formal plan, covering a significant period of time, communicates the scope of the effort and the resources involved. These two factors then point to appropriate monitoring structures and mechanisms. Such issues as deciding between task forces and steering committees, their make-up and frequency/intensity of effort are driven both by the master plan and the dynamics of the business. Whatever the most appropriate mechanism, however, consistency and regularity are the watchwords. The visibility a seemingly high risk or expensive project gets in the beginning often engenders too elaborate a control mechanism. The discipline of maintaining these mechanisms often lags as project implementation goes into operation. Getting the word out both internally and externally maintains the project visibility and the management attention it deserves.

EPILOGUE:
AN AGENDA FOR ACTION

Nathan A. Chiantella
Vice President
Society of Manufacturing Engineers

CIM has been described in this guide from a manager's perspective and as a strategic business opportunity. The thrust is the pursuit of higher efficiencies with faster cycle times as a collective business unit across the entire cycle of product design, manufacture, and marketing.

CIM represents a sizeable challenge to management seeking an agenda for action appropriate to their own product and business arenas. The key question is, how and where to start a CIM initiative?

The answers provided by the authors, who are successfully implementing CIM, are fairly consistent. They have outlined the dependencies—management initiative, stimulation, and participation.

In addition, the authors have characterized the sequence of activities within the four stages of the CIM management cycle. Their emphasis is to focus on management Introduction and Preparation (CIM stages 1 and 2) prior to consideration of a Program Plan and Implementation (CIM stages 3 and 4).

An agenda for action on CIM needs to be sponsored because it is not spontaneous. A place to start is with an "Introduction to CIM;" involve fellow managers. This is critical because CIM is dependent upon building a collaborative effort by the management team. A mechanism to start things rolling is this guide. Pass it on. Use it to stimulate action and to help provide guidance through the four stages of a successful CIM cycle.

INDEX